T0370010

# About Island Press

Since 1984, the nonprofit organization Island Press has been stimulating, shaping, and communicating ideas that are essential for solving environmental problems worldwide. With more than 1,000 titles in print and some 30 new releases each year, we are the nation's leading publisher on environmental issues. We identify innovative thinkers and emerging trends in the environmental field. We work with world-renowned experts and authors to develop cross-disciplinary solutions to environmental challenges.

Island Press designs and executes educational campaigns, in conjunction with our authors, to communicate their critical messages in print, in person, and online using the latest technologies, innovative programs, and the media. Our goal is to reach targeted audiences—scientists, policy makers, environmental advocates, urban planners, the media, and concerned citizens—with information that can be used to create the framework for long-term ecological health and human well-being.

Island Press gratefully acknowledges major support from The Bobolink Foundation, Caldera Foundation, The Curtis and Edith Munson Foundation, The Forrest C. and Frances H. Lattner Foundation, The JPB Foundation, The Kresge Foundation, The Summit Charitable Foundation, Inc., and many other generous organizations and individuals.

Generous support for this publication was provided by Margot and John Ernst.

The opinions expressed in this book are those of the author(s) and do not necessarily reflect the views of our supporters.

# A Poison Like No Other

# A Poison Like No Other

## HOW MICROPLASTICS CORRUPTED OUR PLANET AND OUR BODIES

Matt Simon

Library of Congress Control Number: 2022932929

All Island Press books are printed on environmentally responsible materials.

Manufactured in the United States of America
10 9 8 7 6 5 4 3 2

Keywords: air pollution, bisphenol A (BPA), carbon emissions, climate change, endocrine-disrupting chemicals (EDCS), environmental health, fossil fuels, Great Pacific Garbage Patch, microfibers, nanoplastics, obesogens, plastic bag ban, plastic bottles, plastic pollution, polymers, polystyrene, recycling, single-use plastic, synthetic fabrics, toxic chemicals

To Planet Earth:
Sorry about the mess.

# Contents

# Introduction

On a blazing July day, Janice Brahney leads me up a mountain in the far north of Utah. Brahney is decked out in proper hiking gear and a brimmed straw hat, while I, an indoorsy type, am wearing jeans and tennis shoes. During the winter this is a ski resort, so we pass under chairlifts while I wonder why the operators couldn't run them for just this one summer day. I also live at sea level in San Francisco, so the 1,200-foot climb to the 8,800-foot peak means Brahney spends a good amount of time waiting for me to catch my breath. The wildfire smoke that'd followed me from California, blotting out the surrounding mountain ranges, certainly isn't helping.

I wouldn't be doing this to myself if there weren't something special at the top of Beaver Mountain. As Brahney strolls and I stumble onto the peak, a platform comes into view. On top are two buckets, each containing three layers of sequentially finer screens above a glass plate, which gathers particles falling out of the atmosphere. One bucket is meant to collect depositions only when it rains, so in this awful heat it's covered with a metal plate on stilts. But whenever the sky starts leaking, a sensor detects the moisture and swings the metal plate over to the other bucket, which is meant to collect depositions only when it's dry out. So under different weather conditions Brahney can collect atmospheric grime and, yes, the occasional bird dropping, though that can't penetrate through the screens.

But an insidious new pollutant can: tiny bits of bags, bottles, and other plastics. The nearest city—Logan, where Brahney teaches at Utah State University—is a 45-minute drive away. Yet this humble device on top of Beaver Mountain catches stunning amounts of particles known as microplastics, as do the other instruments that Brahney has positioned around the remote stretches of the American West. Little bits of synthetic material have infested the atmosphere, and they're falling out of the sky.

In 2020, Brahney published an ominously titled paper, "Plastic Rain in Protected Areas of the United States," in the prestigious journal *Science* with a remarkable revelation.[1] Scaling up the number of microplastics she was collecting in these instruments, she calculated that each year the equivalent of 300 million water bottles fall on just 6 percent of the country's land mass. As I put it in a *Wired* magazine story when the paper published: plastic rain is the new acid rain.[2]

Whenever plastic packaging degrades in the environment, it breaks into smaller and smaller pieces. Whenever you wash your polyester or nylon clothes, which are made of soft plastic fibers—two-thirds of clothing is now made of plastic, in fact—millions of threads shed in a single wash and flow to a wastewater treatment facility. From here, the particles are either flushed out to sea or sequestered in "sludge," fertilizer that's then liberally applied to crops. Whenever you drive, chunks of synthetic rubber fly off your tires: each year in the US alone, cars spew 3 billion pounds of the stuff into the environment—that's the wear you see in the treads. Whatever the source of microplastics, and there are many, the particles are accumulating on land, in rivers and lakes, in the sea, and in the atmosphere. "There's no region that's untouched, because the air doesn't care about political boundaries or geographic boundaries," says Brahney, sitting on some rocks and eating blueberries and swatting away bugs as I catch my breath. "Every time I look at my samples, I'm just like, *what the hell, there's so much plastic in here.* It's shocking. And so many different colors, and so many different varieties. I'm just wondering where it originated. And just the sheer amount of it is mind-boggling."

Brahney is part of a vanguard of scientists racing to determine the extent and consequences of microplastic pollution—an invisible, snowballing crisis. While for the past half century environmentalists have been on a crusade against single-use wrappers and bottles, all the while microplastics have spread like a plague around the world, and hardly anyone noticed. If you've wondered about the fate of all that plastic floating in the sea and washing up on beaches and tumbling across the landscape, this is it: mountains of polymers—almost none of it recycled and much of it just chucked into the environment—have broken into an incalculable horde of synthetic specks that will persist for millennia. *Everywhere* scientists look, they find plastic particles, from the depths of the Mariana Trench to the tippy top of Mount Everest and every place in between. Microplastic is the pernicious glitter that has bastardized the whole of Earth, a forever-residue from the party that is consumerism. We have well and truly plasticized this planet, far more thoroughly than images of plastic-clogged harbors would lead you to believe.

You are at this moment exposed to some of the highest concentrations of microplastic anywhere. Stare into the light pouring in through a window and you'll catch glimmers of airborne microplastics flittering around like insects. Leave out a glass of water and you'll find microfibers from your clothes creating tiny dents of surface tension. Leave a glass next to your bed when you change your sheets and you'll see just how many particles the fabric flings into the air. The dust that accumulates in corners and the lint that sticks to your clothes—it's all plastic.

Almost everything around us is spawning microplastics. Carpets are made of synthetic fibers, just like our clothes, and hardwood floors are covered in polymer sealants. Open a bag or bottle and you'll release little plastic shards. Whenever you plop down on the couch, microfibers tear loose and fly into the air. You simply moving around in synthetic clothing—yoga pants, socks, underwear, stretchy denim—creates friction that spawns a billion microfibers a year, according to one estimate. All of it is swirling in indoor air: the floor of a typical living room collects tens of thousands of microplastic particles daily. Your footfalls stir up these particles, resuspending them in the air for you to breathe. Each day we could be inhaling thousands of microplastics and even more *nano*-plastics, which are almost undetectably small and more prone to take to the air. Children, who spend their days crawling through the deposited particles, have it even worse than adults.

This in addition to the microplastic soup that infants are drinking. Scientists have shown that preparing infant formula in a plastic bottle releases several million microplastics into the liquid, what with the heat and vigorous shaking tearing at the material, so babies could be drinking around a billion particles a year. And analyses of adult stool samples suggest we're eating and drinking millions of microplastics a year—indeed, scientists are finding the particles in all manner of foods, from salt to fish. Bottled water is highly contaminated, as you might expect, but microplastics have sullied tap water too. Even groundwater is tainted, the particles having seeped through the soil and into aquifers.

The seas are even more tarnished. One survey of the North Pacific found an average of 8,300 particles per liter of seawater, while a study of the Atlantic estimated that up to 46 billion pounds of microplastics swirl in just the top 650 feet of that ocean. Where currents meet, they aggregate plastic particles, in concentrations many times higher than you'd find in the infamous Great Pacific Garbage Patch. The sea is so lousy with microplastic that the waters are now burping particles into the air to blow onto land in sea breezes. The rest of it eventually sinks to the seafloor: in the Mediterranean, scientists found nearly 200,000 microplastics in a square foot of muck just two inches thick. Other researchers have taken sediment samples to look back in time and show that particle concentrations have been doubling every fifteen years since the 1940s, when plastic production began to accelerate. If humanity doesn't do something to stop the flow of microplastics into the environment, by 2100 there could be fifty times the particles in the ocean as there are today.

As microplastic pollution has continued to grow unchecked, oceanic ecosystems have grown ever more corrupted. The particles are now a fundamental feature of the planktonic community, the very base of the food web. Baby fish are mistaking the particles for food, filling up their bellies and suppressing their appetite for actual sustenance. When bigger fish eat these baby fish, they assume possession of the microplastics, and still bigger fish eat those fish, on up the food chain. Filter-feeding animals like clams and oysters are pulling the particles right out of the water, sequestering microplastics in their tissues for us humans to eat. Scientists have looked in the stomachs of shrimp and marine worms and crustaceans and sea turtles and dolphins and whales, and found microplastics in them all.

The world has never seen a pollutant quite like this. Heavy metals like lead and mercury are but elements, and nasties like DDT are compounds—science knows full well how these pollutants harm life. But the many kinds of plastic polymers contain at least 10,000 different chemicals, a quarter of which scientists consider to be of concern. Plus, as a microplastic particle tumbles through the environment, it accumulates pollutants *already* in the environment, as well as bacteria and viruses, including human pathogens. So microplastic isn't a monolith but a many-headed petrochemical hydra, a plethora of poisons wrapped up in one morsel for Earth's organisms to consume. And being physical chunks or shards or fibers of plastic, these particles physically interact with the environment. Scientists are showing, for instance, that microplastics change the structure of soil, altering how it holds water and how microbial communities form, raising the alarm that the particles could impact crops. Microplastics, then, are a *presence*, more akin to an invasive species—a rat, a kudzu vine, a fire ant—than an ordinary poison, an unprecedented threat to life on this planet.

And they've invaded our own bodies. Autopsies of lung cancer patients have turned up microfibers in their tumors. Doctors have known for decades that people who work in synthetic textile production suffer significantly higher rates of cancers of the lungs and digestive system. Scientists have found microplastics in human blood, colon tissue, and placentas, as well as newborns' first feces, so mothers are passing the particles to their babies. And animal studies have shown that the smallest of plastic particles readily transfer from the gut to the blood to the brain—researchers are confident the same is happening in humans.

What that all means for our health, scientists are just beginning to explore, as the true magnitude of microplastic pollution becomes clear. Of prime concern among the thousands of ingredients in plastic are endocrine-disrupting chemicals, or EDCs, which make hormones go haywire, even in low concentrations. Most infamous among them is bisphenol A, better known as BPA, used extensively in plastic bottles but also linked to depression, sexual dysfunction, and several types of cancer. Phasing out BPA means manufacturers are swapping in chemicals with similar chemical structures that may be just as toxic. And even if BPA were totally phased out tomorrow, we'd still be breathing legacy BPA because long-lasting microplastics have thoroughly corrupted the land, sea, and air. Many of these EDCs are "obesogens," meaning they increase obesity, and toxicologists are gathering evidence that plastics are contributing to the obesity pandemic. Others are investigating if inhaled plastics might also have something to do with rising rates of lung diseases like asthma. They're extra worried about babies, whose bodies are especially susceptible to the hormonal interference of EDCs, and who are gulping millions of microplastics a day in their formula and crawling around in plastic-laden indoor dust: one study found that infant feces is loaded with 10 times the amount of polyethylene terephthalate (a.k.a. polyester) as adult feces.

Humanity has fallen into a plasticine progress trap: the modern world wouldn't be possible without polymers, but the wundermaterials have locked us into an increasingly dark trajectory. Like the invention of agriculture made our species dependent on crops to survive, and like the Industrial Revolution hooked civilization on fossil fuels, so too has plastic set humanity down an ostensibly prosperous path that belies the reality of environmental defilement. Without plastic we'd have no modern medicine or gadgets or wire insulation to keep our homes from burning down. But *with* plastic we've contaminated every corner of Earth and our own bodies, the consequences of which scientists are now desperate to understand.

Extricating humanity from the plastic trap demands an extraordinary—yet feasible—campaign waged across our civilization. There are ways to filter clothing microfibers out of wastewater, to be sure, but technologies are no substitute for what we and the planet really need: a fundamental renegotiation of our relationship with polymers. It wasn't too long ago that humanity got on perfectly fine with cardboard and glass instead of single-use plastic. Plastics *are* fossil fuels, and plastics *are* climate change, so in scorning the material we'll tackle both crises—really, we can't fix one without fixing the other. The production of plastic belches so much carbon, in fact, that if the industry were a country it'd be the fifth-largest emitter, behind China, the US, India, and Russia.[3] And as they age in the environment microplastics release greenhouse gases, suggesting that plastic pollution is a significant overlooked contributor to climate change. Failing to act will doom this world to runaway temperatures and runaway microplastic contamination, the degrees and particles piling up into an unbearable burden for the species of Earth.

CHAPTER 1

# Welcome to Planet Plastic

The road to hell, they say, is paved with good intentions—and a good amount of plastic. The year was 1863, and famous billiard player Michael Phelan was worrying about the sustainability of the very billiard balls that made him a fortune.[1] At the time, the spheres were hand-carved straight out of elephant tusks, ivory being about the toughest material the animal kingdom had to offer. But the things were expensive, and poorly made balls still couldn't withstand repeated smashing without cracking. Also, what if there were suddenly no elephants? Whence would billiard balls come then? Phelan hadn't a clue. But he *did* have $10,000, which he offered as a prize for the inventor who could find a suitable replacement for ivory. Thus Phelan would save the game of billiards and, sure, maybe a few elephants too.

Heeding the call was one John Wesley Hyatt, a 26-year-old journeyman printer.[2] He fiddled with a few different recipes, including a core of wood fiber covered with a mixture of shellac (a resin derived from the excretions of the lac insect) and ivory dust, which was sort of cheating. That and the faux ivory ball didn't have the hardness of the real thing, so billiard players spurned it. Eventually Hyatt began playing around with cellulose nitrate—cotton treated with nitric and sulfuric acids—at his own peril, given that the compound was extremely flammable. Dissolve this cellulose nitrate in alcohol and ether and you get a syrupy solution called collodion, which surgeons used to bind wounds during the Civil War. Hyatt mixed this collodion with camphor (derived from the camphor tree) and found that the product was strong yet moldable. He called it celluloid, and billiard players called it a mixed blessing: celluloid shaped into balls behaved enough like ivory, but being made of cellulose nitrate, they were still . . . fickle. "Consequently," Hyatt later admitted, "a lighted cigar applied would at once result in a serious flame, and occasionally the violent contact of balls would produce a mild explosion like a percussion guncap."

But no matter. Hyatt had invented the first practical, mass-producible plastic, a material that under the right temperature and pressure could be molded into all manner of shapes beyond a sphere. That meant engineers and designers had a new class of material to play with, albeit a volatile one. (Early film was made of celluloid and was therefore super flammable. That's why in Quentin Tarantino's *Inglourious Basterds*, when the good guys burn down the theater with all the Nazis inside, they used a pile of film as an accelerant.) They were no longer stuck tinkering with natural materials like wood and leather, as humans had done for millennia. And glass was a hassle, given its fragility, whereas celluloid was strong yet lightweight. However, though considered a plastic, celluloid was itself largely a natural material, as the cellulose in cellulose nitrate came from cotton and the camphor came from trees—celluloid literally means "cellulose-like," as asteroid means "star-like." (Credit where credit is due: Hyatt had improved on what was technically the first plastic, the cellulose-based Parkesine, which Alexander Parkes never managed to commercialize.[3])

Scientists concocted the first fully synthetic plastic, Bakelite, in 1907. It was borne of the world's shift to electric power, which required insulators for wiring. Shellac did the job, but it was derived from an insect, so manufacturers were limited in the amount of natural material they could procure.[4] By contrast, chemists whipped up the ingredients in Bakelite—phenol and formaldehyde—in the lab. The material kept things from lighting on fire and was durable to boot.[5]

Humans had let the cat out of the plastic bag. Now that scientists knew how to create fully synthetic plastics, and now that the oil and gas business was booming, they could replace natural materials one by one. And the pace of plastic production only accelerated with the material shortages of World War II: nylon replaced cotton, pure rubber was cut with synthetic rubber in tires, and plastic added to glass turned it bulletproof. Read one *Harper's Magazine* article from 1942, headlined "Plastics Come of Age":

> "The Quartermaster Corps is using plastics for canteen closures, the Ordnance Department is using plastics for M52 trench mortar fuses, for pistol grips and bomb detonators, the Navy is using enormous amounts of plastics for insulating electrical parts on ships and in aircraft. Both the Army and the Navy use plastics in almost every form in their aircraft—as fabricated sheets in bomber noses and gun turrets, as molded parts for control boards and instrument housings, as extruded strips for tubing, as foil for electrical insulation, as resin-impregnated plywood or canvas for structural parts, and as resin lacquers for finishing. The fact is that the number of applications of plastics in aircraft construction is increasing so rapidly that the all-plastics plane may precede the all-plastics automobile."[6]

Neither the all-plastics plane nor automobile came to be—your car crumpling and melting around you in an accident is what regulators call "not especially ideal"—but the article did stick the landing more broadly: "It is conceivable that plastics may one day become a dominant material, just as steel did in the immediate past. Or, to put it more strictly (since there are so many kinds of plastics for so many purposes), they may become dominant as metals in general have been dominant from times remote."

To say that WWII hooked the world on plastic like it was an opioid would be an insult to opioids. You can treat a person addicted to a drug, but you can't get plastic out of humanity's system—ever. Being honest, plastic is a miracle material. Get rid of single-use plastics like shopping bags, to be sure, but not plastic syringes and other medical devices, not plastic wiring insulators, not the many components in our cars and electronics. Level any criticism at the petrochemical industry about how they're drowning the world in plastic and the first thing they'll remind you is just how *useful* the stuff is. It's our fault as consumers that we're misusing plastic instead of recycling, which is a bit like opioid manufacturers blaming patients for getting hooked on their drugs.

Like opioids, plastics make everything better in the moment, temporarily masking the ravages of addiction. Just ask the folks jumping for plasticine joy in a two-page spread in the August 1, 1955 issue of *Life Magazine*, "Throwaway Living: Disposable Items Cut Down Household Chores," which must have struck even a vaguely reasonable reader as preposterous.[7] The photo features a radiant nuclear family with arms outstretched, as if worshiping the items falling all around them—plates, cups, utensils, bins, a disposable diaper. "The objects flying through the air in this picture," the story reads, "would take 40 hours to clean—except that no housewife need bother. They are all meant to be thrown away after use." Men need not worry about being left behind in this brave new disposable world, the article hints, thanks to "two items for hunters to throw away: disposable goose and duck decoys." This is the central paradox of plastic: the material is exceedingly valuable in its versatility, yet worthless in that it can be chucked in the bin after one use.

The advertisements on the five pages following the spread are like a staircase leading to the modern consumerist plasticine hellscape. Texaco hypes "that 'cushiony' feeling" of its chassis lubrication. Some sort of living doll with hair made of yarn pours a box of Carnation instant chocolate drink into a glass. "Big-screen color television has arrived!" shouts RCA Victor. A man in a shiny convertible has a problem: he and his kids are enjoying hotdogs, yet deep down he knows that "brushing after meals is best, but it's not always possible." Luckily he brushed before breakfast with Procter and Gamble's Gleem toothpaste, which keeps your maw fresh all day long.

In the decades after *Life* announced the arrival of throwaway living, oil and gas companies like Texaco made the throwaway dream a throwaway reality. A beverage market once cornered by Carnation is now overflowing with brands of soda and energy drinks and juices, all sealed in plastic bottles. The mammoth flat-screen descendants of RCA Victor's 21-inch color TV are made of plastic. Toothpaste isn't just sequestered in plastic tubes—until very recently, it *was* plastic.

In the early 2010s, brands began phasing out the plastic microbeads they'd been adding to toothpaste and face scrubs to boost their scrubbing power. Some of these products contained hundreds of thousands of microplastics, which washed off of your face and out to sea.[8] It turned out that consumers weren't particularly happy when they realized what was happening—President Barack Obama made that displeasure law by signing the Microbead-Free Waters Act in 2015, four decades after microplastic scrubbers were patented in the cosmetics industry.[9] "In that bill, it was only for wash-off cosmetics, and that was mostly the facial scrubs," says Marcus Eriksen, cofounder of the 5 Gyres Institute, a nonprofit that's tackling plastic pollution. "But then in cosmetics, there are tons and tons of shredded microplastic particles used as fillers, things to keep stuff on your face for a long time." Eyeliners, mascaras, lipsticks—they're still loaded with tens of thousands of microplastics each.[10] Microbeads act like ball bearings, making the products more spreadable and silky feeling.[11] By one estimate, over 3 million pounds of microplastics from personal care products still enter the aquatic environment every year.[12] Some 210 trillion microbeads flush out of China alone annually.[13] And while yes, great, the US banned microbeads in wash-off cosmetics, all those particles are still tumbling around the environment and will continue do so for a long, long time.

The microbead battle peaked and waned, and the world patted itself on the back—skirmish against corporations won. But the people didn't know the half of the microplastic problem. Not even *environmental scientists* knew the half of it. Microplastic had by this time become ubiquitous in the environment, and only a small community of researchers had noticed.

Exactly how much plastic humanity has produced thus far, we'll never know. But scientists have taken a swing at an estimate: more than 18 trillion pounds, twice the weight of all the animals living on Earth.[14] Of that, 14 trillion pounds have become waste. Just 9 percent of that waste has been recycled, and 12 percent has been incinerated. The rest has been landfilled or released into the environment, where each bag and bottle and wrapper shatters into millions of microplastics. Sure, many plastic products are relatively long-lasting, like TVs and car components, but 42 percent of plastic has been packaging, very little of which has been recycled.

There's so much plastic pollution out there that if you were to gather it all up and turn it into cling wrap, you'd have more than enough to cover the globe.[15] And this is very much a cling-wrapping in progress: every year, nearly 18 billion pounds of plastic enter just the oceans—one garbage truck full every minute.[16] Just the amount of microplastics entering the environment is the equivalent of every human on Earth walking up to the sea and tossing in a grocery bag every week.[17] In North America, where microplastic emissions are particularly high, it's more like each person contributing three bags a week.

In 1950, when the wide-scale manufacture of plastic was taking off, the industry produced 4.4 billion pounds of resins and synthetic fibers. By 2015, that number had increased almost 200-fold: 838 billion pounds, half of which was single-use plastic—600 million plastic bags are now used every hour, enough to wrap around the planet seven times if you tied them all together.[18] The average American generates almost 300 pounds of plastic waste a year, more than twice that of someone living in the European Union.[19] By 2050, humanity will be churning out over 3 trillion pounds of plastic annually, equivalent to 300 million elephants. That number is all the more stunning when you consider that one of plastic's charms is that it's far lighter than other packaging materials like glass—and it's certainly less dense than an elephant—so you need a whole lot of plastic to reach these weights.

More than half of the plastic ever produced has come in the last two decades, and production is continuing to grow exponentially as Big Oil embraces the inevitable: humanity will someday ditch fossil fuels *as fuels*, but it'll be impossible to ditch the plastic made from fossil fuels.[20] By 2040, the flow of plastic waste into aquatic ecosystems is projected to triple—that means releasing an additional 1.5 trillion pounds of plastic into the environment, and that's a scenario that assumes immediate and drastic action to reduce waste.[21] By the middle of this century humanity will have spent a hundred years producing a total of 75 trillion pounds of plastics and additives, equal to 100,000 Empire State Buildings, at which point four garbage trucks of the material will enter the ocean every minute. And around then, marine plastic will finally outweigh all the fish in the sea.

## Carbon Copies

Any chemist worth their sodium chloride will tell you that carbon is a rather promiscuous element. It likes to form strong, stable bonds with all kinds of other elements: with hydrogen to create methane, with oxygen to create carbon monoxide or carbon dioxide, and—most importantly for petrochemical companies—with itself. Carbon-carbon bonds are the ultra-sturdy backbone of plastic polymers.

Consider the manufacture of polyethylene, which makes plastic bottles. You begin with ethylene, which is two carbon atoms—stuck together with a double bond—each of which has two hydrogen atoms attached.[22] To turn this into a plastic polymer, chemists break the bond between the carbon atoms, allowing them to connect more carbon atoms in a chain. This is the *poly*mer (derived from the Greek for "many parts") made up of *mono*mers (from the Greek for "alone"). (Polymers actually abound in the natural world: wood and rubber are polymers, as is the tough shell of a beetle.)

Here are the most common plastic polymers, along with their uses. As major players in the microplastics crisis, they'll pop up throughout this book.

- Low-density polyethylene, a.k.a. LDPE: single-use shopping bags and food packaging like cling wrap and the lining of take-out coffee cups.
- High-density polyethylene, a.k.a. HDPE: the tougher sibling of LDPE, found in bottles and bins.
- Polyethylene terephthalate, a.k.a. PET: makes for good water bottles. The polyester in your synthetic clothing is also made from PET.
- Polyvinyl chloride, a.k.a. PVC: most famously used in piping, but also in packaging and in plastic children's toys.
- Polyurethane, a.k.a. PU: commonly used as a varnish.
- Polystyrene: better known to consumers by the brand name Styrofoam.
- Polypropylene: a rigid plastic that can be steam-sterilized, hence its use in medical equipment.

What differentiates these plastic polymers are the atoms attached to that carbon chain. These are known as side chains, and they make a particular polymer more waterproof, for instance, or more resistant to heat. Different polymers also come in different densities, the obvious examples being low- and high-density polyethylene, which determines if they're more prone to sink or float in water. But whatever the polymer, it's built on the strength of carbon.

All this carbon has to come from somewhere, and that's cheap, plentiful fossil fuels. "When you've done the extraction of the oil and the refining of the oil, all the things that are left over then are basically used as the basis of pretty much all of our synthetic chemicals," says chemist Iseult Lynch of the University of Birmingham, who studies plastics. "So it's actually much, much cheaper just to keep refining virgin plastic than it is to recycle or reuse or recover. And that's the main problem, that if they're extracting it for the oil and gas industry, then these byproducts are always going to just be cheap." This is why humanity has only recycled 9 percent of the 14 trillion pounds of plastic waste it has produced: it's cheaper to just churn out more virgin plastic than it is to reprocess what's already out there. The economics of recycling aren't just busted—they're preposterous.

Every step of manufacturing virgin plastic also expels greenhouse gases. It takes energy to drill into the earth and extract fossil fuels, plus wells leak methane, a greenhouse gas that's 80 times more potent than carbon dioxide. (In fairness to methane, it also disappears from the atmosphere much quicker than $CO_2$, on a timescale of decades, not centuries.) Transporting fossil fuels takes energy. Turning them into monomers to make polymers takes energy: according to the Center for International Environmental Law, 24 ethylene facilities in the US emit as much $CO_2$ in a year as 4 million passenger vehicles.[23] The group further calculates that if the production and use of plastic continues to grow on its current trajectory, by 2030 the industry will generate as much greenhouse gas per year as 295 coal plants. By 2050, that will more than double to 615 coal plants. Plastics production is accelerating so feverishly that the industry's emissions will overtake emissions from coal by 2030, according to the advocacy group Beyond Plastics.[24] So really, plastics are canceling out the gains humanity has been making on climate change by decommissioning coal plants and electrifying vehicles. "The fossil fuel industry does not want to give up sales, so they are now making a very big play on plastic production," says Judith Enck, president of Beyond Plastics and a former Environmental Protection Agency regional administrator. "It's rather alarming, and it's happening under the radar."

Between 2019 and 2021 alone, at least 42 new plastics facilities either came online, were under construction, or were in the permitting phase. Which is not to say that you the consumer are what's driving this boom in plastic production—no one's asking for more of their stuff to be wrapped in single-use plastic. Petrochemical companies and food and beverage giants are manufacturing the need for plastic because it boosts their bottom line, as plastic containers are lighter than glass and therefore cheaper to ship. If large-scale recycling actually worked, the industry would stop building plastics facilities, not build more of them. "The petrochemical industry is still trying to fool people into thinking that plastics recycling is a solution, and it's been an abysmal failure," says Enck. "We are dealing with both the fossil fuel industry *and* the chemical industry, neither of which wants to make less plastic."

In addition to the emissions from the production of plastic, the carbon within the material doesn't stay there forever. Sarah-Jeanne Royer, a biogeochemist at Hawaii Pacific University, has been experimenting with how different plastic polymers emit greenhouse gases and found that polyethylene, the most common plastic, releases the most methane.[25] More ominous still, Royer found that polyethylene in powder form produced 488 times more methane than in a larger pellet form.

The reason is counterintuitive. As a piece of plastic fragments into microplastics, growing tinier and tinier, it *increases* its relative surface area. Think about potatoes. If you thoroughly microwave a big one, it'll take longer for the center to cool to edibility, whereas a smaller potato will cool to its center quicker. The big potato has more volume to trap that heat, whereas the small potato has relatively more surface area to lose that heat. The same goes for microplastics: the smaller they get, the more surface area they have relative to volume. Meaning, more of the microplastic material is exposed to the medium it's tumbling around in, be it soil or air or water.

So say you have a small sphere of plastic. The material at its center is locked away from the environment, but cleave it in two, and more surface area opens up. Cleave each of the two halves again, and you get more surface area still. You'd do the same to cool down that microwaved potato quicker. Except for our purposes, instead of steam escaping from a potato, it's greenhouse gases escaping from microplastics. "Which means that as plastic degrades in the environment," says Royer, "a plastic bottle becoming smaller and smaller in size is just going to produce more and more greenhouse gases, and in an exponential form."

Royer ran these experiments on polymers in different mediums and found that plastics exposed to air emitted twice the methane and 76 times the ethylene (another greenhouse gas) as those in water, which is troubling given that Janice Brahney and other scientists are finding that the atmosphere is loaded with microplastics. Exactly how much planet-warming gas the petroparticles are emitting while they're up there, and also while they're floating in the sea and tumbling over land, Royer can't yet say. It'd be a doozy of a calculation: you'd have to know how much of the stuff is out there, and the proportions of different polymers, and how big the particles are, just to name a few variables. But for now we can say that microplastics may well be a significant—and significantly understudied—source of greenhouse gas emissions.

But what about plastics not made from fossil fuels? You may have recently opted for bio-based plastic of some type, maybe for your compost or dog poop bags. These are made of monomers derived from plants like corn or sugar instead of fossil fuels. You reckon that you can fill up a bio-based bag, toss it in the compost bin, and the collection company will add it to a great big pile of waste, where the bag will simply melt away. But yeah, no, not necessarily, and tests have shown that even the ones that do readily break down release greenhouse gases in the process.[26] Bio-based doesn't always mean it's biodegradable, and biodegradable plastics aren't always bio-based.[27] That and "biodegradable" doesn't always mean it's the kind of biodegradable you're expecting—oftentimes a compostable bag can only break down quickly in an industrial facility, where the polymer simmers in temperatures well over 100 degrees Fahrenheit.[28] "*Biodegradable* is one of those funny words that, from a regulator's point of view, means that it degrades under a specific set of time at a specific set of predefined conditions," says Lynch. "Those conditions might well be: at a certain temperature and under a certain set of microbial conditions. So it doesn't actually mean what you or I, as a consumer, would think, *Oh, we're buying something biodegradable, that means it's like a banana skin that we can just throw away*. Even a banana skin degrades over a period of months."

That and even if a biodegradable bag *did* dissolve in soil, that's one environmental condition it was designed for. If it got out into the ocean, that's an entirely different set of circumstances—temperature, salinity, acidity, microbial communities—which it's ill-designed to interact with. Funny enough, one experiment found that a compostable bag buried in soil for over two years came out mostly whole, though it did tear when scientists tried to carry things in it.[29] But when exposed to seawater for just three months, the same kind of bag disappeared. So the composting process would actually work better if you threw your scraps in the ocean, which probably isn't what the manufacturer had in mind. Another "biodegradable" bag the researchers tested held up so well after three years in seawater that you could still use it to carry home your groceries—if you didn't mind the smell of the marine life that'd grown on it.

Biodegradable just means the plastic breaks down faster—theoretically at least. It'll still be fragmenting into microplastics: one study found that an ounce of biodegradable plastic left in soil breaks into 60,000 particles.[30] And plenty of other research has shown that bio-based microplastics are just as toxic as regular microplastics because they contain the same noxious additives that make plastic plastic.[31] It's not that the biodegradable bag isn't still out there—it's just been deconstructed, like a star that's exploded and flung its component parts out into the universe, except microplastics are amalgamations of petrochemicals, any one of which may be toxic to a particular species.

One group of researchers is calling this the "global plastic toxicity debt." An unfathomable amount of plastic is already out there and will be out there for a long time—breaking into smaller and smaller pieces, leaching more chemicals, coming into contact with more species. "Given the production patterns, which have been steeply increasing," the researchers write, "and the time scales of degradation with estimated half-lives ranging from less than a year to several thousand years depending on the plastic and its chemistry, we very likely have not reached the peak of toxic release from the sum total of all environmentally available plastic on Earth. This is a sobering thought."[32] Like racking up credit card debt, we're pumping plastic into the environment, racking up a toxicity debt, "but that has many more years of decay and release of toxic compounds to follow." Even if we were to cut up the credit card and stop the flow of plastic into the environment, the debt will remain out there, splintering and splintering and splintering into ever smaller bits.[33]

If plastics were just made of carbon chains, that debt may be benign. After all, life on Earth is carbon based. But monomers are complex molecules, and polymers are more complex still. Some polymer chains, for instance, will line up neatly in rows, making them brittle. So chemists have to add chemicals known as plasticizers, which rest in between the layered polymer chains, pushing them apart. This makes plastics malleable, yet strong. But these plasticizers turn out to be extraordinarily toxic—the most common variety, phthalates, cause severe hormone problems in humans.[34] Chemists also throw in flame retardants, antioxidants, and UV-absorbing stabilizers, which slow the degradation of polymers in the sun but also pile on the global plastic toxicity debt.

All told, plastics manufacturers have used thousands upon thousands of chemicals in their products. The Nordic Council of Ministers, an intergovernmental group, has identified 144 individual chemicals or chemical groups used in plastics that are known to be hazardous, including formaldehyde and toxic metals like cadmium and lead.[35] (Metal-based additives act as fillers in plastic, thus increasing stiffness and hardness, and as antimicrobials, flame-retardants, and pigments.[36]) Only a quarter of the 4,000 chemicals approved for use in food packaging in the United States have been evaluated for safety.[37] "One of the hard parts is that companies won't tell us what they put in our plastic," says Duke University environmental engineer Imari Walker, who studies microplastics. "So it creates a very difficult problem for us to know what we're looking for."

Adding yet another layer of complexity, Walker and other scientists are finding that these chemicals aren't static: they can transmogrify over the life of the plastic. One study found that when exposed to sunlight, a single-use bag from Walmart spit out 15,000 compounds distinct from the compounds original to the plastic.[38] "What you originally put in there can change when you expose it to a beach setting, when you expose it to light, wind, air," Walker says. "Those chemicals can transform. And the question is, are they more or less harmful when they're doing those transformations? So you go from having the original hundreds, thousands, sometimes maybe just 10, and they can end up breaking apart into who knows how many different combinations once in the environment. And so how do we look for them all? And how do we know they're safe?"

For as strong as these polymer chains are, they do indeed break down as the plastic tumbles around the environment. A single-use plastic bag floating in the ocean is bombarded by the sun's UV radiation, which creates free radicals in the presence of oxygen. Over time, these free radicals break apart the polymer chains, releasing the plasticizers and returning the plastic to a brittle state. "All the additives, they're only just sort of physically entrapped in between the polymer chains," says Lynch. "If you think about when you make jelly, the water just gets trapped in as it solidifies, but it's not chemically bonded in there." Once the polymer chain begins to fragment, that jelly deconstructs, and out come the additives. This leachate, as environmental scientists call it, is a complex chemical cocktail that doses whatever living things happen to be in the vicinity.

As the brittle plastic bag floats across the sea, getting flung about by waves, the mechanical action breaks it apart. Temperature changes, especially freezing and thawing, further assault the structural integrity of the plastic. Organisms like algae colonize the synthetic raft, weighing it down. As it sinks, the plastic escapes UV bombardment, but now it has to contend with pressure changes. Currents pick it up and whip it around. Where those currents meet, plastics accumulate, bumping up against one another like celluloid billiard balls. Polymer chains break apart, over and over and over, and the plastic deconstructs into smaller and smaller pieces. Bits that you can hold in the palm of your hand. Then bits that fit on your thumbnail. Then bits you can just barely make out with your naked eye. And finally they go microscopic, disappearing from human view to make trouble for some other poor critter out in the sea.

Determining how fast plastics break down, though, is difficult, given all the different polymers and different conditions out there.[39] What scientists can manage at this point is re-creating natural conditions in the lab. At the University of Bayreuth in Germany, chemist Nora Meides sets microplastics loose in a weathering chamber, in which a xenon lamp blasts the particles with UV radiation.[40] About 3,000 hours in here equals 15,000 hours in the wild, since this xenon "sun" never sets. Meides has found that in the first 600 hours in the chamber, a plastic particle's size decreases linearly as its surface cracks. But after that, the decay increases exponentially as the particle splits into new pieces. "We have breaking apart—fresh surface is exposed, where surface degradation starts again," says Meides. "But from the decrease in size, we have also a faster weathering, basically. So we have an exponential decay in size from that stage on." That is, as a microplastic splits into new pieces—smaller potatoes, if you will—more surface area is exposed to UV bombardment, decaying the particle until it splits into more pieces, and those pieces split into more pieces.

But this experiment doesn't speak for all the plastics undergoing unique trials out in the environment. Ocean temperatures vary widely, as do pressures. Plastics floating at the surface are subjected to UV bombardment, while denser polymers that sink will live in darkness. On land, plastic faces a different battery of stressors—more abrasion against surfaces, for example. When they get small enough, microplastics take to the air, confronting still more unique conditions. Microplastics readily move between these realms—living out distinct experiences like a human transitioning between childhood, adulthood, and senescence—as they contaminate every corner of Earth.

"If you've got an IQ above room temperature," says Steve Allen, a microplastics researcher at the Ocean Frontier Institute, "you have to understand that this is not a good material to have in the environment."

## Sizing Up Microplastics

By scientific definition, a piece of plastic becomes a microplastic when it gets smaller than five millimeters, about the diameter of a pencil eraser.[41] Experts proposed the definition during a 2008 meeting of the National

Oceanic and Atmospheric Administration, in which they reached the following point of agreement: "Microplastics as defined here (<5mm) are not likely to cause widespread ingestion-related effects on large organisms (e.g., birds, marine mammals), certainly relative to the well-documented impacts of larger marine plastics."[42] Put another way: we're all familiar with plastic bags ending up in turtle stomachs, but what smaller plastics might smaller organisms be eating? If a particle is 5.1 millimeters long, you'd hate to rob it of the microplastic honorific, but for the standardization of microplastic science, it's necessary for researchers to agree on a number when they're collecting particles, otherwise they wouldn't be able to compare their findings. So five millimeters it is.

(A quick procedural note. Scientists speak in the metric system, but throughout this book I've translated figures into American measurements for an American audience—kilograms to pounds, meters to feet, and so on. Researchers measure water volumes in liters, so I'll stick to that because—and I say this as an American with a practiced sense of irony and self-loathing—I assume my compatriots know how much a liter is on account of two-liter soda bottles. The five-millimeter figure will come up here and there because it's a tidy number compared to the equivalent of 0.19685 inches, so just keep in mind that five millimeters is the diameter of a pencil eraser.)

Because the study of microplastics is such a new science, researchers are still codifying the standard methodologies for both gathering and characterizing the petroparticles. The first consideration will be size. Two research teams studying microplastics in the ocean might use different filters when sieving out particles: if you took the same liter of water and passed it through filters of different fineness, you'd get different particle counts, as smaller bits will slip through a coarser filter. And once you count the particles in the lab, the tallies can diverge even further, as one team might have the technology to count the tiniest of particles while the other settles for a less thorough count.

It's also possible to overcount microplastics. The particles have so thoroughly saturated our surroundings that microplastics researchers need to take special precautions not to contaminate their samples. So if a scientist gathers an air sample from a city and brings it back to the lab, they need to be sure that once they open the container, fibers floating in the indoor air don't sneak in. They might take care to eliminate as much plastic as possible from the lab—and certainly using plastic containers to do sampling is a big no-no—and wear only cotton clothing. But how do you tell a white synthetic microfiber from a white cotton fiber? It wasn't that long ago that microplastics researchers would—and this is no joke—jab particles with hot needles to see if they'd melt or curl up like a synthetic material would. If it smelled funny, it was probably plastic.

Microplastics researchers have since developed much more sophisticated methods for determining the composition of a suspected microplastic: Fourier transform infrared spectroscopy (FTIR) and Raman spectroscopy. FTIR works by firing a laser at a particle and detecting the infrared signature that bounces back. The system then compares this unique signature to a database of known materials. Not only does it finger a particle as plastic, but the signature will also say what type of polymer it is—polyethylene will look different to the machine than PVC. (Polymer chemists used this technique before microplastics science became a field.) Raman spectroscopy works along the same lines, only it fires a laser at the particle and detects how the chemical bonds scatter the light at different wavelengths. Like with FTIR, the system compares this signature to a database of known materials, thus identifying a particle as a specific polymer.

As the technology has improved, scientists have been able to detect smaller and smaller particles. Get down below five millimeters, below one millimeter, and much farther still, and you enter the nano realm: once a speck of plastic gets smaller than a micrometer, or a millionth of a meter, a microplastic becomes a nanoplastic. (Though scientists are still debating what the exact micro/nano threshold should be.) Not only are nanoplastics far too small for your human eye to see—an individual bacterial cell is between 1 and 10 micrometers long—they're too small for all but the most sophisticated (and expensive) of imaging equipment. Plus it's hard to collect the stuff in the first place: nanoplastics will slip right through a net fine enough to snag microplastics. Nanoplastics research, then, is just getting off the ground. Scientists have detected plenty of the particles in the environment but don't yet have a good idea of the quantities in the land, sea, and air.[43] But if the saturation of microplastics out there is any indication, the nanoplastic numbers will be even more astronomical: a speck of microplastic will fragment into a horde of individual nanoplastics.[44] Indeed, early research in the remote Alps found that 19 billion nanoplastics fall on a single square foot of land each week, suggesting the atmosphere is packed with the particles.

But scientists still don't know how small plastics can get as they degrade in the environment. Do they eventually break down into their component chemicals? Or do the particles get so small that they reach a kind of sweet spot, where they're no longer subject to abrasion? Do they then persist in the air or ocean as clouds of invisible synthetic material? "I think by the time we get down below 100 nanometers, the material itself has virtually no kinetic energy, so it's going to be very difficult to break," says Steve Allen. "And the wavelengths of light are generally longer than the size of the particle, so it's going to have no effect on the particle at that point." However small plastics get, researchers are extremely concerned about the human health impacts of nanoplastics, which can likely move through human organs, the brain included. They're even small enough to enter our cells.

Even quantifying microplastics is hard as hell. The depressing reality is that their numbers are increasing so rapidly—both because we're releasing ever more into the environment year over year and because as microplastics degrade they split into more pieces, like proliferating bacteria—that a sample site can be more contaminated when a scientist publishes their findings than it was when they visited just a year or two prior. "We're constantly underestimating," says microplastics researcher Deonie Allen of the University of Strathclyde. (Deonie and Steve Allen are plastic-hunting spouses.) "Pretty much all of our results are immediately outdated. We're constantly fighting an uphill battle to try to keep up."

I'll say this loud, because it'll come up time and time again throughout this book: *a given tally of microplastic particles in water, soil, or air is almost certainly a significant undercount—but some undercounts are better than others.* When I say, for instance, that researchers found 500 particles in an ounce of soil, I mean they found 500 particles that were big enough to detect with their methods. The smaller the researchers can go, the better their figure represents reality: as the particles decrease in size they increase in number, so a sample can contain millions of microplastics and trillions of nanoplastics.[45] But because nanoplastics remain difficult to detect, great quantities of the smallest of bits are evading detection.

Just because nanoplastics are exceedingly small doesn't mean they're harmless. The opposite may prove true. As a plastic particle's edges whittle away over time, more untouched material at the center is exposed, which then leaches its chemicals. At the same time, microplastics and nanoplastics are accumulating pollutants from their surroundings. Scientists have found that toxic metals like mercury and lead readily glom on to microplastics, concentrating at levels hundreds of times higher than the medium the particle is tumbling through. The same goes for a bevy of pesticides like DDT, as well as pharmaceuticals, including antibiotics, pain relievers, and benzodiazepines like Ativan.[46] Organisms also quickly colonize a microplastic particle: microbes like viruses and bacteria, many of them human pathogens, grow alongside algae and even tiny larvae of animals. It's an ecosystem so rich and so distinctive, scientists have dubbed it the plastisphere.

## Nurdles, Fibers, and Wet Wipes

Microplastics can flow into rivers and oceans already tiny, in the case of microbeads, or they spawn from plastic litter. But microplastics are also escaping into the environment *before* they become proper plastics.[47] The raw material that makes up a plastic product is known colloquially as a nurdle, a hilarious word that the industry eschews for "pellet." Nurdles are between two and five millimeters and come in different shapes, but what they have in common is that they're easy to lose. "It's like trying to hold a handful of sand," says Andrew Wunderley, executive director of Charleston Waterkeeper, a watchdog group in South Carolina. "It's kind of impossible to hold them in your hand without spilling them all over the place. So anytime you're moving them, you're up for spills."

In 2021, the pellet transport company Frontier Logistics agreed to pay $1 million to settle a suit brought by Charleston Waterkeeper and the Coastal Conservation League alleging that one of its facilities was leaking nurdles.[48] The nurdle pollution problem is especially acute along the coasts of Texas and Louisiana, epicenters for oil and gas and plastic production. Following a pellet spill in the Mississippi River in the summer of 2020, volunteers collected 250,000 nurdles on a 10-minute walk down a Louisiana beach, according to Jace Tunnell, who runs Nurdle Patrol, a citizen science project. "They were just filling up five-gallon buckets," says Tunnell, who's also the director of the Mission-Aransas National Estuarine Research Reserve at the University of Texas. Previously, in 2019 the petrochemical company Formosa Plastics agreed to pay $50 million to settle a lawsuit accusing it of dumping billions of pellets into Texas waterways,[49] the largest settlement of a Clean Water Act suit ever brought by private citizens.[50]

Nurdles escape at any link in the supply chain. They spill when they're loaded into containers at a production facility, and when they're loaded onto ships or railcars. Then they fall out of those ships and railcars: in 2012, a typhoon knocked containers off a vessel near Hong Kong, spilling 330,000 pounds of nurdles, which gathered on beaches like snow.[51] In 2021, a ship cruising through the Indian Ocean caught fire and sank, losing 87 containers full of nearly 4 million pounds of nurdles—the largest plastic spill ever—which piled six feet high on the Sri Lankan coastline.[52] Upping the toxicity of the spill, scientists discovered that many of the nurdles had burned in the wreck, making them three times as chemically complex as virgin plastic.[53]

Nurdle spills *are* oil spills, essentially, only they travel much farther. "These things are tiny, they're lightweight, the wind blows them around," says Tunnell. "If it happens to fall on the ground, they're considered contaminated, so they can't just suck them up and put them back into the vessel that they were being transported in. They have to be either thrown away or recycled. The product is so cheap that it's more expensive to pay somebody to clean it up than it is to leave it on the ground." Rains wash the nurdles into rivers, where the less dense polymers like polyethylene float and eventually flow out to sea.[54] Denser polymers like PVC will sink to the bottom of a river, belying the true extent of the problem: just because you can't see a nurdle spill doesn't mean it isn't there.

One estimate reckons that worldwide each year, 500 million pounds of nurdles enter the oceans.[55] In a Swedish harbor outside a polyethylene production plant, nets pulled up 380 microplastics per gallon of water.[56] Another survey of the beaches of São Paulo, Brazil, found a galaxy of nurdles—which likely came from a nearby port—up to six feet deep.[57] In 1994, researchers found as many as 82,000 nurdles per square foot of sand in a Jordan nature reserve, also from spills.[58] Up to 53 billion pellets could be entering the ocean from just the UK each year, the equivalent of 35 tanker loads.[59] And because polymers can be less dense than seawater, the nurdles travel far from industrial centers, washing up on the most remote of beaches. Beachcombers—folks who search the world's sands for valuables—call the translucent beads "mermaid tears."[60]

(Let me pause here and say that this stuff is heavy, I know it, and I'm feeling your unease. But I promise we'll be talking about solutions, ranging from individual actions to ways to break plastic's grip on our civilization. Microplastics are a serious problem, but not an impossible problem to mitigate. We'll be meeting very smart people who are working on it.)

Now, microplastics other than nurdles come in two varieties: microplastics proper, and the subspecies microfibers. A microplastic might be thick or thin, long or stout. It might be spherical or lumpy or square or triangular, or something entirely more irregular. As it abrades over time, its corners might round off. Maybe it'll split in two, or just slough off wee bits of itself as the years pass.[61] Plastic wrapping breaks down into films or flakes. Styrofoam crumbles like feta. "There's no such thing as *a* microplastic," says Steve Allen. "They're all completely different." The world doesn't have a microplastic problem—it has a microplastics problem, plural.

Microfibers tend to be more homogenous. Whenever you wash your underwear or socks or yoga pants—plastic imparts stretchiness, in addition to water- and stain-proofing—tiny fibers break off in the laundry water. These microfibers come in every color you can imagine an article of clothing would come in and can be any length up to five millimeters. As they age their ends may fray, like split ends of hair, and cracks or pits may form along their lengths. But generally speaking, microfibers are . . . fibrous, whereas other microplastics come in a broader range of forms.

But what microfibers lack in morphological diversity, they make up for in numbers. The microfiber-rich water from your laundry machine flows to a wastewater treatment facility, where it's pumped through screens to remove large debris, then sent to a "grit chamber," in which sand and stones settle to the bottom.[62] The water then moves to a sedimentation tank, where the human waste and other organic material settle out, and then to another tank in which microbes further break down organic material left in the water. Finally, chlorine is added to kill any remaining nefarious bacteria, and the (relatively) clean water is pumped out to a river, lake, or ocean. Wastewater treatment facilities aren't specifically equipped to filter out microfibers, but just by accident through all these processes, they'll capture anywhere between 83 and 99 percent of the tiny pollutants, depending on the equipment.

But we're flushing so many microfibers to these facilities that even a single percent of them escaping into the environment is a huge problem. Between 1950 and 2010, the production of synthetic fibers exploded from 4.6 billion pounds to 110 billion pounds a year. Two out of three pieces of clothing are now made of plastic, textiles like polyester, nylon, and fleece.[63] This coincides with the rise of fast fashion, crappy clothes that quickly wear out and disintegrate, shedding fibers at a rapid clip. More than half of all fast fashion is disposed of within a year, according to one estimate,[64] and a garbage truck full of textiles is now landfilled or incinerated every second.[65] It's a vicious circle: synthetic clothing keeps humans warm on the cheap, and as more people rise into the middle class, they demand bigger wardrobes—total annual clothing sales are expected to triple by 2050. We just don't have the land and water to grow the cotton to meet this demand, but we do have plenty of fossil fuels.

Even if we could go back to wearing clothes made from all-natural fibers like wool and cotton, those aren't blameless here either, because they're no longer all-natural. "It's not as simple as just snapping your fingers and removing all synthetic clothing because even natural clothing can have synthetic polymers associated with them," says Lisa Erdle, manager of research and innovation at the 5 Gyres Institute. Most machine-washable wool, Erdle notes, is coated in a layer of polyurethane. Cotton is treated with synthetic dyes and other chemicals, including flame retardants, UV stabilizers, and antimicrobial agents—the same additives in plastics.[66] These chemicals may make up a third of a "natural" garment's weight, Erdle says. "Even those natural materials do have elements of a plastic in them. It's not as simple as just banning a single material."

You might be wondering why we can't just slap filters on washing machines, like we have lint traps on our dryers. It's a good question with a maddening answer: washing machines in North America did have filters at one point. "If you look at older machines—and I recently rented an apartment that had an old washing machine from the late '80s—there's a lint trap," says Erdle. Manufacturers of washing machines can build in filters, but they don't, because so many of us have dryers with filters instead. Erdle's research shows just how effective it'd be in keeping microfibers out of the environment: an aftermarket filter captured 87 percent of particles.[67]

In the same study, Erdle's team set out to quantify how many fibers are shed in each wash, so they got themselves some polyester fleece blankets from IKEA (the Polarvide, if you were curious), washed them in a top-loading machine without detergent, and collected the drain water. They found a single load generates between 91,000 and 138,000 microfibers. They then took that figure and extrapolated how many fibers the city of Toronto—which has 1.2 million households, each doing an average of 219 loads a year—would be flushing to wastewater treatment plants annually: up to 36 trillion. (They had data from a single kind of IKEA blanket, so obviously this isn't representative of the diversity of clothing the good people of Toronto launder.) Even if those facilities capture 99 percent of the particles, the remaining 360 billion would still be pumped out to sea each year—from a single, moderately sized city.

Over in the UK, another team of researchers ran a more complicated experiment with duplicates of three kinds of sweaters and found that a load of clothes sheds 138,000 polyester-cotton fibers, 500,000 polyester fibers, and 730,000 acrylic fibers.[68] Let's pretend you've got equal proportions of the three fabrics in your wardrobe and average those fibers per load to 456,000. If you're doing 219 loads a year, that's 100 million microplastics flowing from your house to a wastewater treatment plant. Even if this is a best-of-the-best facility that captures 99 percent of microfibers, you'd still be flushing a million fibers out to sea annually.

This study also found that detergent increased the number of fibers shed, which other scientists in Sweden also found.[69] This could be due to detergents decreasing surface tension, which facilitates wetting of the fibers in clothing to chase out grime—and loose fibers. Detergents also work as dispersing agents: the soap dissolves dirt and keeps it suspended in the wash water, so the grime doesn't settle back into the clothing. Detergents may do the same with microfibers, freeing them from clothes and keeping them locked in the wastewater that flows to a treatment facility. Still another study found that liquid detergent increased the shedding of fibers 8-fold and powder detergent 22-fold compared to washing in water alone.[70] That may be due to the grittiness of the powder scouring clothes—almost like sandblasting—increasing friction between synthetic fibers and releasing them en masse. All told, these scientists landed on an even higher figure of over 6 million polyester fibers shed per load. Doing our quick calculation again: if you washed 219 loads a year, and a wastewater treatment facility caught 99 percent of your microfibers, that would mean your household would send 13 million particles to the environment each year.

So wait, is a load of laundry generating tens of thousands, hundreds of thousands, or millions of microfibers? Well, these studies differ because washing clothes is complicated business. We have many different washing machines to choose from, and we have our preferred wash cycles—regular, delicates, permanent press—which liberate more or less microfibers. We use a multitude of detergent brands in liquid form, powder, or packs. We choose between hot and cold. All of these preferences become *scientists'* preferences in the lab, plus they're washing unique articles of clothing.

All of these studies, though, are undercounts, because they haven't been considering nanoplastics. In 2021, scientists published the first study to quantify the nanoplastics shed from polyester in a washing machine, finding that one load releases millions of microfibers, but also hundreds of trillions of nanoplastics.[71] "So the question is, What are actually the processes producing them, if it's not UV degradation?" asks Bernd Nowack, an environmental scientist at the Swiss Federal Laboratories for Materials Science and Technology, who did the experiment. "We don't know whether we *produced* them—at least we *released* them." Meaning the nanoplastics may have been left over from the manufacturing process, and a wash flushed them out of the fabric. This would explain why other studies have found that with repeated washing, clothes release fewer microfibers—the number never gets to zero, but it appears that new garments shed the loose fibers they were born with. Which is actually a spot of good news in the microplastics crisis: if we buy more quality clothes and wear them for years and wash them as little as possible, we'll emit fewer microfibers than if we buy fast fashion, which quickly disintegrates into particles. With well-made clothes you'll have to buy fewer replacements that shake off the loose fibers left over from manufacturing.

At the moment, though, households all over the world are cooking up a kind of plasticine soup, which then flows into rivers and lakes and oceans. One Glasgow wastewater treatment plant serving 650,000 people was found to release 65 million microplastics a day.[72] Three facilities discharging to South Carolina's Charleston Harbor—two of which could remove 85 percent of microplastics, the other 98 percent—each day collectively emit up to a billion particles.[73] And keep in mind that we're talking about developed countries here with proper wastewater infrastructure: if you'll pardon the mental image, consider that worldwide each year, humans produce enough wastewater to fill 144 million Olympic-sized swimming pools,[74] and only half of that is ever treated.[75]

One study estimated that between 1950 and 2016, over 6 billion pounds of microfibers have escaped our clothes and entered water bodies, equivalent to over 7 billion fleece jackets by mass.[76] Half of those emissions have come in just the last decade, and by 2050, washing machines will be churning out 1.5 billion pounds of plastic a year. It's no wonder, then, that when scientists sample seawater or a beach, they find that clothing microfibers dominate, often hundreds of them in a footprint in the sand.

So we need microfiber filters on every washing machine that rolls off the line from here on out—France is leading the way on this, mandating that all new models come with filters by 2025—and in the meantime, governments should distribute aftermarket filters to their citizens as a matter of environmental emergency.[77] The things cost as little as $45, and there are around 100 million washing machines in the US, which would work out to a bill of $4.5 billion. One study found that if Californians all adopted washing machine filters, they'd reduce the state's microfiber emissions by 79 percent.[78] There's the issue of disposal, though: if you toss your captured microfibers in the trash, they may well take to the air during the waste management process. So a company called PlanetCare makes a filter with cartridges that you ship back so they can dispose of the fibers—they're considering turning it into building insulation.

But even before all these clothes become clothes, they're shedding microfibers. From plastic pellets manufacturers make synthetic yarn, which becomes synthetic fabric, which becomes synthetic clothing. Along the way, "wet processes" like dyeing and washing in great big agitating machines create microfiber-laced wastewater, which each year loads the environment with 265 million pounds of microplastics, according to an investigation by the Nature Conservancy. For every 500 T-shirts made, a full one escapes to the sea as microfibers. If that pollution is left unchecked, by 2030 the figure could go up 50 percent, on account of the cancerous growth of the fast fashion industry, yet another reason to buy quality clothes that don't need so much replacing.[79]

The fashion industry has at least begun researching ways to prevent textiles from shedding microfibers. In 2018, a nonprofit called the Microfibre Consortium formed to facilitate a response by a range of actors in fashion, from manufacturers to brands to retailers. It's got around 70 signatories so far, including Nike, Gap, and Target, with a goal of more than tripling that number by 2030.[80] The consortium has developed standardized guidelines for testing how fabrics shed microfibers, so a brand can quantify how one of its products performs, then share that data anonymously with the consortium, which compiles reports on the state of the research across the industry. Using an online hub, retailers, manufacturers, and policymakers can access this data—basically, instead of everyone researching new fabrics in isolation, companies can tap into what others are discovering. "It's that data dissemination platform where people can go in and they can get very detailed and science-led information with regards to any product changes they want to make," says Sophie Mather, executive director of the Microfibre Consortium. "They could get information about different yarns, different fabrics, and how they need to communicate that to their suppliers with regards to developing new materials."

One approach could be filament yarns, which are made of longer, continuous strands that seem to be less prone to shedding. Compare this to a polyester fleece, which is brushed in the manufacturing process to raise its surface, making it softer and fluffier, and therefore better able to trap air and keep you warm. But that scouring obviously generates microfibers: one test of 37 different textiles found a huge range of emissions in a wash, from a few thousand microfibers per garment to many millions, with treated fleece releasing on average six times the particles as nylon woven with filament yarns.[81] (The microfiber champion, at 4 million particles released, was a "soft double velour recycled fleece.") Keep in mind, though, that this lab, and any other lab in the Microfibre Consortium, is not a home: laundry is complicated business—temperature, detergents, agitation—so a garment that tests well for scientists might still release microfibers in your washing machine. And let's not forget the hundreds of trillions of nanoplastics that come out in a single wash. Still, the fact that researchers are finding massive variability in the microfibers shed from different clothes gives hope that the industry can shift to those low shedders.

But people *like* fleece. It's warm and soft and cuddly. A brand can do the right thing for the environment and switch to filament yarn, easy enough, but that could be the wrong thing for their bottom line. "So there is a lot of balancing to be done in the material development teams to really understand where can the change be made to stop or to reduce fiber fragmentation without compromising what is important to them as a product," says Mather. "Because the last thing you want is a product that is not fit for purpose, and therefore becomes obsolete and waste."

If one brand stops making microfiber-spewing fleece, you can be damn sure another will step in to meet demand, especially as the fast fashion industry swells. There's no law that governs microfiber pollution, so we're at the mercy of the brands that created the problem to engineer solutions to fix the problem. Perhaps it's time, then, to use the market *and* regulation to slow the flow of microfibers into the environment: label clothing to tell consumers how much each garment sheds. "So some kind of traffic-light system, where green means it doesn't shed very much, amber means it sheds a bit, and red is that you're going to be introducing a lot of particles into your home by buying this," says Fay Couceiro, who studies environmental pollution at the University of Portsmouth. "If we have that kind of knowledge in society, theoretically I would like to think that would bring down the amount of plastic."

It'd be even easier to fix the other big contributor of wastewater microplastics: bathroom wipes are also made of plastic or other synthetic fibers like rayon, a kind of processed cellulose. Same goes for baby wipes and sanitary pads. You're not likely to find any mention of synthetic material on the label, but it is indeed what keeps the wet wipes from disintegrating in the package before you use them. It's also what keeps them from disintegrating once they leave your toilet.

As the popularity of "flushable" wipes took off in the mid-2000s, the folks who attend to our sewage started complaining of uber-clogs in their systems. In 2019, a presumably exasperated Municipal Enforcement Sewer Use Group of Canada commissioned a report on flushable goods, things like tissues, paper towels, wipes for both baby and adult bottoms, and diaper liners.[82] Engineers at Ryerson University tested 101 products in a "slosh box," which tilted back and forth 11 degrees (the researchers were very clear about this) for a half hour, agitating each product in 59-degree-Fahrenheit water, thus mimicking a journey through the sewer. Presumably adding to the exasperation of the Municipal Enforcement Sewer Use Group of Canada, the engineers found that only 17 of the products showed evidence of disintegration. And of those 17 products, 11 fully disintegrated—because they were all bath tissues. Conversely, two of the 23 products explicitly labeled as flushable partially disintegrated, while the rest didn't deteriorate a bit because they were synthetic, containing rayon, polyester, or polypropylene.

To test whether these ostensibly flushable products are shedding microplastics as they move through a sewer system, scientists sampled sediments along the coast of Ireland, comparing microfiber counts from a site near a wastewater treatment plant to two control sites, one 10 miles away and the other 62 miles.[83] They also bought different brands of wet wipes and sanitary towels and analyzed their fibers, finding that the most common polymer was polyethylene terephthalate, or PET. These products were all white, so when the researchers sifted through the sediment samples they could seek out that color, then confirm that the environmental fibers' chemical signatures were the same.

In addition to finding all kinds of intact wipes and sanitary pads that'd washed ashore at the main sample site, they found 78 white fibers per ounce of coastal sediment, which represented 91 percent of the total microplastic particles in the sample. By contrast, at the nearest control site they found only 11 fibers per ounce, and still fewer at the more distant site. The white fibers all matched virgin fibers taken from the products in the lab, and in fact didn't show much weathering at all. Wipes shedding fibers in a sewage system, then, work a lot like clothes shedding fibers in the wash: the material *seems* to resist breaking down, but in fact it's sloughing off little bits of itself that wash out to the environment. Ban them, then—ban anything meant to be flushed that's made of plastic—because second by second, the oceans are growing more and more infested with microplastics.

CHAPTER 2

# A Voyage on the Synthetic Seas

In the fall of 1971, budding oceanographer Edward Carpenter sailed through the Sargasso Sea, far off the Atlantic Coast. From the research vessel he dragged neuston tows—fine nets that skim along the surface—to collect the sea's namesake sargassum, a frilly kind of seaweed. In the first few tows he got a whole lot of sargassum, as expected, but also noticed a smattering of tiny plastics. Nothing to write home about, he thought—just an aberration. But then another scoop brought up more particles. "Then all of a sudden, it hit me by the third neuston tow that the plastics are ubiquitous," Carpenter recalls. "And they're out here about as far away from land as you could be. I was really shocked."

When Carpenter got back to land, he typed up a short paper and shipped it off to the journal *Science*, which published the findings on March 17, 1972—one of the first documentations of microplastics in the environment.[1] (Though Carpenter didn't call them such, as the term wasn't coined until 2004.[2]) Most of the particles, he noted, were white, but some were blue or green or red or clear. Many were brittle, indicating that the plasticizers within had been lost to weathering. "The increasing production of plastics," Carpenter wrote in the paper, presciently enough, "combined with present waste-disposal practices, will probably lead to greater concentrations on the sea surface."

At least one industry association, the Society of the Plastics Industry, took note. (It's now called the Plastics Industry Association, and it did not respond to a request to comment for this book. The American Chemistry Council, the other major plastics industry group, agreed to field questions but then did not respond.) "They sent a guy out to talk with me," says Carpenter. "Basically, he came out to pretty much intimidate me, and sort of belittle the research a bit." Carpenter can't remember the man's exact words, but he recalls them being generally not encouraging. "I know most all of the time, he had a scowl on his face."

Carpenter paid the man no mind. Later that year he published a second paper in *Science* describing how he'd collected yet more plastic spherules in the coastal waters of New England.[3] This time, Carpenter also gathered 14 fish species and found that 8 of them had consumed the particles. "If you took the size of those little fish, it was like we'd have a bowling ball or a watermelon in our stomach," says Carpenter. "So you know that the fish weren't going to be able to pass anything that big through their digestive system."

At the end of the second paper, Carpenter acknowledged that industry rep: "We thank R. Harding for his cooperation and for notifying all polystyrene manufacturers in the United States of the presence of plastic spherules in coastal waters." The industry knew it had a microplastic problem, and Carpenter's discovery got a fair amount of press, including a writeup in the *New York Times*, so it was on the public's radar.[4] But attention in the '70s instead increasingly focused on *macro*plastic pollution like bags and bottles, and Carpenter drifted away from plastics, instead building a career studying other aspects of ocean biology. It wouldn't be for another 30 years that the study of microplastics got going in earnest. And in those 30 years, the Sargasso Sea and every other bit of the ocean was growing ever more infested with tiny bits of plastic.

Carpenter had stumbled into a grand gathering of migratory microplastics. The Sargasso Sea is surrounded by major ocean currents that form a clockwise swirl in the Atlantic: catch the current in the north and you can ride east to Europe, while the southern current will take you back to the Americas. Such currents are the transoceanic highways on which ships—and microplastics—ride.

The most obvious way to generate a current is with wind, which scours the sea surface and drags water along. This creates turbulence, so a piece of floating plastic is liable to get mixed down, at least temporarily. "If you put a bunch of plastics in a cup of water, and then you shake it up, the plastic will disperse everywhere," says Michelle DiBenedetto, who studies the transport of microplastics at the University of Washington. "But then over time, it'll flow back to the surface. So when you have winds, that wind is basically mixing up the surface—it's putting energy into the ocean." This energy sets the top layer of water in motion, but that water also tugs a bit on the water below it. Something, though, has to replace the water that's now moving out of the area, which is how you get upwelling, as liquid from below rushes up to fill the void. So wind triggers both a lateral and vertical movement of seawater.

The other way to form an ocean current is by playing with the density of water. If a current flows north into the Arctic, for instance, the surface water may freeze, but the salt in it won't, instead sinking into deeper water. This pushes down on the waters below, squeezing them out of the Arctic toward the equator. Currents that flow along the seafloor also slow down or speed up depending on the topography, while surface currents interact with shorelines.

Throw in tidal motion, and you're looking at an ocean that's constantly churning, dragging microplastics far and wide, from populated coastlines to the most remote of seas.[5] One survey of the Atlantic calculated that up to 46 billion pounds of microplastics are suspended in just the top 650 feet of that ocean.[6] The average depth of the Atlantic is 12,000 feet, so that's leaving out a lot of water, and that survey only included microplastic fragments, not microfibers. On the other side of the US, a sampling of the waters of the Monterey Bay—a famous conservation success story—also turned up loads of particles,[7] while another survey of the North Pacific, which looked for particularly small microplastics, found an average of 8,300 per liter of seawater.[8] We can expect different regions of the sea to vary in their microplastic pollution, given factors like currents and proximity to human populations, but a synthesis of over 8,000 samples collected by various scientific teams across the world's oceans between 2000 and 2019 estimated that 24 trillion microplastics float just at the surface of the sea.[9] Second by second, that number is rising, so any tally like this is immediately outdated. Like astronomers can only estimate how many stars are in the sky, so too can oceanographers only estimate how many microplastics are in the sea—except the universe is mostly done making new stars, and humanity has just begun making microplastics.

Oodles of particles are washing ashore too. A sampling of six beaches on Oahu found up to 160 microplastics per square foot of sand.[10] They're on the beaches of remote Easter Island and the Galapagos too.[11] In the sands of even-more-remote Henderson Island, smack dab between New Zealand and South America, scientists found over 400 pieces of plastic debris per square foot.[12] They also discovered that the particles are absorbing the sun's energy and raising the temperature of the sand by more than four degrees Fahrenheit.[13] This could have two huge consequences. One, sea turtles lay their eggs in beach sand, where the temperature determines the sex of the hatchling (hotter temperatures make more females). Biologists are already seeing clutches turn out nearly all female because of climate change,[14] and nesting grounds the world over are now loaded with heat-absorbing microplastics.[15] And two, by increasing surface temperatures, microplastics may be helping heat parts of the planet. So as more and more petroparticles flow into the environment and shed from oceanic macroplastics, beaches don't just get more polluted—they get hotter.

Microplastics are also accumulating in seafloor sediments. Think of the seafloor like a countryside of hills and valleys through which rivers flow. On land, sandbanks grow where rivers bend and the water slows, losing its energy and allowing sediment to fall out and accumulate. On the bottom of the sea, the uneven topology directs ocean currents in the same way—where those currents slow, they also drop their suspended particles, which build up over time as sediment. So at the bottom of the Mediterranean Sea, scientists found way more microplastics where currents were slower: 176,000 in a square foot of muck just two inches thick.[16] The vast majority of the particles were fibers, and in fact some samples came back 100 percent microfiber.[17]

The culprit is wastewater. The Mediterranean Sea is ringed by 29,000 miles of coastline, home to 150 million people.[18] Think back to that estimate of Toronto's contribution to microfiber pollution: 360 billion particles coming from a city of 3 million people. That study was assuming that mere thousands of microfibers go out in a wash—when follow-up studies found it's more likely to be millions, not even counting nanoplastics—and that wastewater treatment removes 99 percent of particles. But scaling up that very conservative estimate to the population of the Mediterranean coast gives us 18 trillion microfibers entering the sea each year. Obviously that's rough math with a whole lot of variables in play—methods of clothes washing, wastewater treatment technologies, and so on. But that's a best-case estimate in which facilities capture as much of the plastic as they're able. Now consider that a fifth of the Mediterranean's coastal cities lack those facilities altogether.[19] That and the sea exchanges little water with the Atlantic through the Strait of Gibraltar—what goes into the Mediterranean, mostly stays in the Mediterranean. Accordingly, the sea is now more polluted with microplastics than the Great Pacific Garbage Patch, where converging currents accumulate trash.[20] (The name makes it sound like an island of solid garbage you could walk across, but the stuff is more spread out than that, plus the vast majority of it is invisible microplastics, not macroplastics.)

Clothing fibers are ubiquitous—and I mean *ubiquitous*—in the world's waters and sediments, but they're far from the only brand of microplastic pollution out there. Fishing vessels often lose their massive plastic nets and other gear, to the tune of 100 million pounds a year worldwide.[21] All this "ghost" netting bakes in the sun and abrades against other netting, and against itself, sloughing off microplastics.[22] When crews drag synthetic nets through the water or across a seabed for miles and miles, microfibers tear loose, just as clothes shed in a washing machine.[23] Plastic traps and pots do the same. When crews haul their equipment back aboard, that friction generates still more fibers: one experiment showed that rope releases up to 230 particles per foot—a 500-foot rope, then, would produce 116,000 particles every time a ship pulled it up. The researchers estimated that each year, the 4,500 active fishing vessels in the UK alone could be emitting up to 17 billion microplastics, and that's assuming a modest rope length of 160 feet.[24] That and they calculated this in their experiment by pulling just five and a half pounds, so they caution that their estimates are conservative: ships are hauling aboard much longer ropes with much heavier loads and creating more friction and more microplastic.

In some parts of the world's oceans, plastics coming off ships beat out fibers as the most abundant microplastic. Little bits of marine paint—a thin coat of plastic designed to keep life from attaching to ships—leave "skid marks" that contain plasticizers, heavy metals like lead, and flame retardants.[25] "These microplastics can be really rich in biocides, the chemicals that are designed to stop animals and plants growing on the surface of boats," says Matthew Cole, a marine ecologist and ecotoxicologist at the Plymouth Marine Laboratory in the UK. "So it's not just the traditional plastics that scientists are concerned with—there's lots of different forms." Scientists are finding that in harbors in particular, staggering amounts of paint particles are accumulating in sediment and in local fish species.[26] Not helping matters is the fact that back at port, crews remove paint by blasting it with abrasive microbeads—a double release of microplastic.

Oceanographers studying the waters around Germany and Denmark, which are crisscrossed by some of the world's busiest shipping lanes, found that these paint plastics can outnumber other oceanic microplastics four to one. Some 5,500 container ships are currently in operation,[27] and ocean shipping is growing, full steam ahead: over the last three decades, total trade volume has tripled, reaching 12 billion tons in 2018.[28] And that's to say nothing of all the other ships scooting around the oceans—yachts, oil tankers, military vessels—as well as bridges and piers and oil rigs, all of which shed microplastics like dandruff. Scientists on a research cruise in the remote Weddell Sea off Antarctica discovered that half of microplastics in the waters they sampled were actually paint fragments, many from their own ship.[29] And any vessel releases sewage, known as black water, and the wastewater from showers and sinks and laundry, known as gray water. Cruise ships alone could be releasing a quarter of a billion pounds of microplastics via gray water every year, according to one estimate.[30]

Plus, there's all the paint on land, countless structures baking in the sun and molting microplastics. It'd be difficult to estimate how many particles fly off buildings each year—consider all the variables, like the type of paint and the climate—but we can safely assume that the contribution is substantial: some scientists reckon that a third of the microplastics escaping into the environment may be paint particles.[31] Even though lead has been phased out of new paints, all this erosion of old paint means that the heavy metal still makes its way into ecosystems via microplastics.

Paint polymers, ghost net particles, and microfibers are conspiring to create serious ecological consequences. When currents deposit microplastics in the undersea equivalent of sandbanks, they're also depositing organic matter. So the petroparticles land where they can do the most damage: where food amasses, so does the life that depends on it. Creatures like sea stars and sea urchins roam this "benthic" zone, munching on the organic fallout—and now microplastic—while stationary species like sponges snag food out of currents. Animals like sea cucumbers, which hoover up the sediment itself and filter out food, now get mouthfuls of synthetic material too. When predators consume these prey, they're also consuming the microplastics within.

And they've been doing so since the mid-20th century. Core samples of sediments off the coast of Southern California show that between 1945 and 2009, microplastic concentrations doubled every 15 years.[32] If you plot the sedimentary microplastic counts for each year on a graph, the curve overlaps perfectly with the exponential acceleration of global plastic production over the last eight decades. "It will become our fossil record," says oceanographer Jennifer Brandon, who led the research. "And the reason plastic is the perfect marker to look for is that plastic lasts forever. It doesn't really degrade. It already basically *is* fossilized." Other scientists have found a similar signal in sediments around Asia and in mangrove swamps along the Saudi Arabian coast.[33] Worldwide, there may be 32 billion pounds of microplastic in the top few inches of sediment, and that's an extremely conservative estimate based on samples taken from the deep sea, not coastal areas, which are more polluted on account of their proximity to human populations.[34]

How exactly the petroparticles will preserve for our descendants to find, no one can yet say. In seafloor sediments, microplastics might last indefinitely, given the cold temperatures and that they're not bombarded by UV light. Even if plastic does degrade in the sediment, it might leave a "cast" as the material around it solidifies, a cavity that says *Plastic Was Here*. "Thus," wrote geologists in a 2016 paper, "the outlines of biros, plastic bottles or compact disks (CDs) may be found as fossils in sedimentary rock in the future even if the plastic itself has degraded or been replaced by other materials."[35] It'd be a new, shameful kind of stone built of pollution. "Plastics are already present in sufficient numbers to be considered as one of the most important types of 'technofossil' that will form a permanent record of human presence on Earth."

Not even the remote Arctic has been spared. With some clever sleuthing, scientists figured out how microfibers not only enter the Arctic Ocean but get stuck there.[36] This ocean has two main inputs: water from the Atlantic flows between Greenland and Norway, and to a lesser extent, some water comes in from the Pacific, squeezing between Alaska and Russia. These researchers sampled seawater along a transect across the Arctic, from a few feet deep to 3,300 feet, and found all of it teeming with microplastics. Over 90 percent of the particles were microfibers, three-quarters of which were polyester in a range of colors—obvious remnants of synthetic clothing.[37] The scientists used FTIR to confirm the type of plastic, which also provided clues as to the age of the microfibers, which in turn provided clues as to how the particles are moving around the Arctic Ocean. The team found three times as many fibers around the Atlantic input as they did around the Pacific input, but microfibers were three times longer in the former than the latter, suggesting they hadn't had time to weather and break apart. In addition, the FTIR signature of the Atlantic particles more closely matched that of virgin synthetic fibers.

So, putting that all together: Wastewater plants along the coasts of the US, Canada, and Europe are releasing microfibers in effluent. The particles enter the Atlantic Ocean, whose currents carry the microfibers into the Arctic. The same is probably happening along the coasts of the Pacific, but a smaller input to the Arctic Ocean there means fewer of those fibers reach the northern waters. That's why the scientists found shorter fibers here: they're not young fibers that floated up from coastal populations around the Pacific, but weathered fibers originally from the Atlantic. They'd swirled around the Arctic Ocean like flakes in a great big snow globe, aging and fragmenting while making their way toward Alaska.

It's not just wholly synthetic fibers that scientists are finding in the Arctic. As Lisa Erdle of the 5 Gyres Institute pointed out, even cotton is treated with dyes and other chemicals—this is known as anthropogenically modified cellulose. Erdle has sorted through muck from the deep Arctic Ocean, the Great Lakes, and shallow suburban lakes around Toronto and found 875, 353, and 1,130 microfibers per pound of sediment, respectively.[38] Up to half of the microfibers were anthropogenically modified cellulose, and of that, half were denim (indigo gave them away).[39] But the consequences of this pollution will be unique for each body of water. The Great Lakes receive almost 5 billion gallons of wastewater a day but let almost none of it go because they're essentially a closed system, so the fibers are accumulating in the water, sediment, and animals.[40] "The quantities that we're finding of microfibers in these fish are often an order of magnitude higher than some of the highest reported concentrations for the ocean, even in fish in the middle of the Great Pacific Garbage Patch," says Erdle. "And that's just because the Great Lakes are so concentrated." The Arctic Ocean, by contrast, is a more open system: currents constantly shuttle water in and out.

But that doesn't mean microfibers don't cause trouble up there. At first, marine biologist Melanie Bergmann, of the Alfred Wegener Institute in Germany, worried that it'd be difficult to detect microplastics in sediment samples she gathered during Arctic Ocean expeditions—surely she'd need to collect a lot of muck to find a signal. But one early sample came back with 2,700 particles per pound of sediment, and more than twice that in another. "We needn't have worried about the amount of sediment that we need to sample to find anything," says Bergmann. "While everyone presumes that the Arctic is such a nice, pristine place, it's not."

Also contrary to popular belief, the Arctic Ocean isn't devoid of life. What is muck to you and me is food for seafloor scavengers, which eke out a living filtering sediment for what little organic material falls from the water column above. It's exactly *because* there's less biomass on the Arctic seafloor than, say, a coral reef that microplastics may be a particular problem there. "It's costing already food-deprived organisms more energy to process this, without any energetic gain or very little energetic gain," says Bergmann. A sea cucumber gets a mouthful of synthetic material, and—best-case scenario—nothing to show for it. Worst case, the microplastics are harming it.

The deluge of microplastics compounds the troubles that Arctic organisms are already struggling to surmount. The region is warming four times as fast as the rest of the planet,[41] and it's caught in a vicious feedback loop. Being white, ice reflects the sun's energy—its "albedo" is high. But as more ice melts, it exposes the darker seas underneath, which have a lower albedo and absorb more energy, further heating the waters and further heating the region. At the same time, oceans are acidifying due to the chemical reaction of excess atmospheric carbon dioxide with the water. And now Arctic animals are eating microplastics. "Often studies just look at the effect of ingestion," Bergmann says, "but not at the effects of ingestion and chemicals; or ingestion and temperature change; or ingestion, temperature change, and ocean acidification. But the animals in the environment, of course, experience all these different effects."

Plus, Bergmann and other scientists are concerned that microplastics *themselves* may be changing the albedo of the Arctic. They're finding the stuff sprinkled across the ice, and if enough dark-colored particles amass here, they'll absorb more of the sun's energy—just like microplastics are warming up beaches—and accelerate melting. And the plastic is embedding in the ice itself: as crystals form at the ocean surface, they "scavenge" particulate matter like microplastics from the water.[42] Bergmann has found 340,000 microplastics in a cubic foot of Arctic ice.[43] Given rates of melting, that'd mean northern ice releases an astounding number of petroparticles each year—8 followed by 20 zeroes. Some of the microplastics will resume snow-globing around the Arctic Ocean, freeze once more, then melt again, and on and on, while others will catch currents south toward the coastal cities whence they came.

### Plastic on the Menu

Apparently the only things less steady than my hiking-up-an-8,800-foot-mountain legs are my sea legs. It's a tumultuous November morning on the deck of the research vessel *Rachel Carson*—named after the famed biologist, writer, and conservationist who alerted the world to the peril of pesticides in her classic book *Silent Spring*—and I'm trying my best to keep an eye on a horizon erased by fog. We're an hour off the coast of California, where scientists from the Monterey Bay Aquarium Research Institute have lowered an SUV-sized robot 1,600 feet deep. They're in the control room piloting the thing. I'm out here and the Dramamine isn't helping, but every half hour or so I stabilize enough to rush back inside, observe, and rush back out again.

The live feed coming from the remotely operated vehicle is just too magical to resist. Cruising ever so gently below us, the robot seems to fly through a galaxy, only the stars here are tiny bits of organic matter, known as marine snow. Every so often some little organism darts away like a meteor, spooked by the ROV's light and whirring motors—hard to blame it really, since on any other day this is a kingdom of darkness. The robot cuts through clouds of free-swimming crustaceans and gelatinous critters called ctenophores, whose beating hairlike structures flash a rainbow of color in the robot's beam.

We've entered the realm of the giant larvacean. It's not so much the tadpole-shaped animal itself that's giant—it only grows to a few inches long—but the three-foot-wide house it builds out of mucus. By beating its tail, the larvacean drives up to 80 liters of water each hour through its house, which traps larger bits of organic matter that wouldn't fit in its mouth while the smaller stuff slips through. When the net grows saturated with gunk, the animal jettisons it and builds a new one. Here at 1,600 feet deep is a whole city of these abandoned homes, great suspended gobs speckled with detritus that shine in the light of the ROV. Given enough time, they'll get weighed down and start sinking half a mile a day, providing a nicely packaged buffet for scavengers in the depths. "Sinkers," scientists call them.

Four years prior to this nauseating journey, biological oceanographer Anela Choy was on the same ship using the same robot to show that microplastics are going along for the ride into the deep sea, the biggest habitat on Earth. Science still knows vanishingly little about these depths, but it does know this: they're now loaded with petroparticles, thanks in large part to the sinking houses of giant larvaceans. Polluted surface waters are intimately linked to the deep, even though in parts of the ocean, the realms are tens of thousands of feet apart. "There would be no life in the deep sea if it were not for what is happening at the surface ocean," Choy says. "You can think of microplastics as a component now of that."

Aboard the *Rachel Carson* in 2017, Choy used the ROV to slurp up giant larvaceans at different depths, then tested them for microplastics. Sure enough, the larvaceans had particles both in their guts and sprinkled in their houses, in concentrations that matched the microplastic pollution in the water at that particular depth. In a separate experiment, Choy released microplastics of various sizes around giant larvaceans in the wild and watched as the larger particles got stuck in the animals' houses while the smaller ones made it through.[44] After collecting some of the specimens and adding them to a holding tank aboard the ship, Choy watched as the larvaceans pooped out the particles.[45] But more worryingly for the denizens of the deep, these experiments show that giant larvaceans the world over are snagging microplastics in their houses, which all kinds of deep-sea creatures eat indiscriminately. "Plastic ingestion within the food web is potentially as complex as the food web itself, but it's also very prevalent," says Choy. "The amount and the quality of that organic matter is really what dictates life in the deep sea."

Every day, ocean life embarks on a mass migration that puts a bird flock or reindeer herd to shame. When the sun's up, animals large and small hang out in the relative safety of deeper, darker waters, where their predators can't see them. But at night, they move en masse to surface waters, where there's more sustenance—and more microplastics they mistake for sustenance. The extreme diversity of microplastics, all shapes and sizes and colors, provides a diversity of different-looking "food" for different foragers with different diets.[46] Indeed, a 2009 expedition in the Pacific Ocean collected fishes and found microplastic in the stomachs (they didn't look in the rest of the digestive system) of 4.8 percent of species that don't do this kind of vertical migration, as it's known, while 11.6 percent of migrating species tested positive for plastic.[47] This suggests that the migrants are picking up more particles closer to the surface when they feed, whereas the non-migrating fishes feed exclusively in depths, where there may be fewer microplastics. All told, these researchers calculated that in the gyre that forms the Great Pacific Garbage Patch, fishes consume up to 53 million pounds of plastic debris each year just in the mesopelagic zone, between 650 and 3,300 feet deep. Migratory species are therefore acting as vehicles to disperse microplastics farther down the water column: when they return to the darkness of the depths and poop, out come the particles. The fecal matter and microplastics join all the other organic debris from giant larvacean houses and decaying organisms to make marine snow, which falls to the seafloor and accumulates as sediment.

So even in the deepest depths, ocean life is feeding on microplastics. Between 2008 and 2017, oceanographers visited nine sites, ranging from 23,000 feet to 36,000 feet at the Challenger Deep in the Mariana Trench (the deepest point in the ocean), and collected amphipods, crustaceans that look like shrimp.[48] At every site, at least half of the animals had microplastic in their guts. The Mariana Trench's amphipods had *all* ingested microplastic, and they tallied the highest average particle count per individual among the nine sites. The vast majority of these particles were fibers. A separate expedition found up to 2,200 microplastics per liter of sediment from the Mariana Trench, so it's no wonder the animals down there are so contaminated.[49]

In another survey in the middle of the Atlantic, three-quarters of captured deep-sea fishes had microplastics in their bellies.[50] One specimen of a common fangtooth had two prey in its stomach, a cock-eyed squid and a bearded sea devil, which themselves had microplastics in their guts. (As a general but informal rule, scientists tend to give deep-sea animals names that are at once fantastical and perfectly descriptive. The fangtooth has a mouth full of giant teeth, one of the cock-eyed squid's peepers is enlarged to collect faint light from above, and the bearded sea devil grows frilly bioluminescent structures on its chin.) This reveals a dynamic called trophic transfer: a small species eats microplastic, then a predator eats that small species, and a still larger predator eats that predator[51]—a perpetual biological cycling of microplastics. If you were to somehow instantly remove all the particles from ocean waters and sediments, they'd live on by transferring from gut to gut. Everything eats, everything gets eaten, and microplastics go along for the ride.

Microplastics aren't just in the food chain—they've penetrated the very *base* of the food chain. Surface waters teem with little organisms collectively known as plankton, which are divided into two groups. We've got bacteria and algae known as phytoplankton, which harvest their energy from the sun. Though individual phytoplankton are microscopic, great blooms of them create green clouds that drift along the surface of the world's oceans. Like plants do on land, these photosynthetic organisms take in the carbon dioxide dissolved in seawater and spit out oxygen, and a whole lot of it: they're responsible for two-thirds of the atmosphere's oxygen content.[52] The carbon the organisms absorb goes the other way, sinking to the seafloor when the phytoplankton die. These life forms, then, are critical actors in the carbon cycle, sequestering $CO_2$ and locking it in the ocean depths.

They're also a critical source of food for the other group of plankton: zooplankton. These include tiny species of animals like crustaceans and jellyfish and marine worms. To reproduce in the open ocean, a female fish releases masses of eggs and a male releases sperm, which all drift around in planktonic clouds. When those eggs hatch into larval fish, they become more active participants in the community, eating phytoplankton and each other. They are, in turn, eaten by bigger fish and seabirds.

Now you can add microplastic as a third player to the vast planktonic universe sparkling across Earth's oceans. Scientists have analyzed a trove of stashed-away plankton samples, collected off the coast of Scotland since the 1960s, and found a significant rise in microfiber contamination over the decades—a sad echo of Jennifer Brandon's sediment samples.[53] Others have taken water samples off Antarctica and found zooplankton tangled up in microfibers.[54] Biologists not only find petroplankton and biological plankton mingling together in water samples taken from different depths but also find the particles mingling in the stomachs of captured zooplankton: baby salmon that (eventually) feed bears and humans, the crustaceans that feed fish and birds, the krill that feed whales.[55] A survey in the South China Sea tested fish larvae, jellyfish, shrimp, and predatory worms and found microplastics in them all.[56] Adult fish that feed on zooplankton, or on smaller fish that feed on zooplankton, then assume the particles: a sampling in the remote South Pacific turned up microplastics in 97 percent of fish species, including mahi-mahi, red snapper, and barracuda[57]—one foot-long Pacific chub was burdened with 104 pieces of plastic in its gut. It's worth noting that the most contaminated fish were in the most remote waters, around Easter Island, where the South Pacific Gyre—the counterpart of the

North Pacific Gyre, which creates the Great Pacific Garbage Patch—accumulates plastic.

Any size and shape and color of microplastic you can dream of, it's out there, so a speck too big for one hunter becomes easy pickings for the next. Nurdles, which look just like nutritious fish eggs, may be particularly tempting. Not many marine animals are big enough to choke on a whole plastic bag, but a great many diminutive creatures can choke on microplastic they mistake for prey: lab experiments have shown that some zooplankton prefer eating aged plastics over pristine plastics by a wide margin.[58] These animals may be sniffing out the microbes they typically eat, only to instead ingest petroplankton with microbial frosting.

The plastisphere, as scientists Erik Zettler and Linda Amaral-Zettler are calling it, is a new kind of microbial habitat on Earth. Moving from surface waters to deep waters and back again, from the Atlantic to the Arctic, from coral reefs to open ocean, microplastics gather miniature communities like stamps on a passport.[59] "Basically it's a mini-world," says Linda. "There are organisms that make their own food, that photosynthesize. There are predators there. There are prey. There are symbionts and there are pathogens as well—or at least as we know them to be pathogens on animals, and potentially humans as well." Under a microscope, this mini-world bustles.[60] Wispy filaments, possibly produced by bacteria, wrap around kayak-shaped photosynthetic organisms known as diatoms.[61] A single-celled organism called a ciliate grows like a mushroom from the surface. Keep zooming in and you'll see it's got company: the unmistakable tubular forms of bacteria cover its bulb but—more curious still—not its stalk.

Far from being a sanitary piece of synthetic material, each microplastic crawls with life. "It's this three-dimensional structure that provides all kinds of different niches for primary producers, consumers, degraders, parasites, predators, grazers—you name it," says Erik. Bacteria and viruses mingle with animals, like the larvae of barnacles. "Each of those animals has its own microbiome of additional microbes that are brought to this. So it really becomes quite a complicated little community."[62] As these microplastics tumble around the sea, they fragment—some organisms will hold on, others won't. Microbes may themselves break down the plastic: Linda and Erik have found spherical cells resting in pits on a microplastic's surface, suggesting that whatever species this is, it could be digesting the polymer. "At some point," Erik says, "when you get down to the nanoplastics range, it's not going to be the microbe sticking to the plastic, it's going to start becoming the plastic sticking to the microbes."

Erik and Linda have only begun to understand this world. They can test the DNA on microplastics to reveal what kinds of bacteria are present, but that can't tell them how those players are interacting with one another. The bacteria *Vibrio*, for example, has dominated many of their samples, and other researchers have found the microbe hitching rides on microplastics elsewhere, from the Baltic Sea to the coasts of China and Brazil.[63] This is the bug in undercooked seafood that can cause severe gastrointestinal distress, but Erik and Linda caution that just because *Vibrio* is there, often in significant quantities, doesn't necessarily mean it's a threat to the health of sea creatures or humans. What's clear, though, is that ocean life has never encountered anything quite like the plastisphere.

Consider the journey of a polyester microfiber. During its life on your sweater, it'll have gathered microbes from your skin and the air around you. Flushed to a wastewater treatment facility, the fiber is soaked in human waste, acquiring new microbes there.[64] In a river, the plastisphere steeps in agricultural runoff rich in nutrients, which opportunistic microbes might thrive on.[65] Finally in the open ocean, the particle encounters a still more alien saltwater microbiome—bacteria, viruses, animal larvae. Floating at the surface, it attracts sunlight-loving organisms,[66] but then grows so heavy with life that it starts sinking.[67] In the darkness, the microbes dependent on light will perish, happily replaced by others in the new habitat.[68] As currents carry it dozens and hundreds and thousands of miles, our fiber will sample innumerable unique oceanic microbiomes, its own microbiomes reshuffling all the while.[69] It may even act as a shuttle to introduce microbes to new oceanic habitats.

And then, disaster—mass extinction in the plastisphere. A baby crustacean hoovers up the microplastic and its hangers-on, which are now imprisoned in a hostile habitat meant to digest things. Much microbial death later, the fiber comes out the other end. "There may be a few really hardy survivors that are still left on basically pristine plastic, and it gets put out into the ocean with a little dollop of fertilizer as well," Erik Zettler says. "So all of a sudden you have this yummy piece of pristine surface with high nutrient concentration available to a whole new community. That might be very different from the community that formed when that plastic first went into the ocean."

Zooplankton that feed on algae could instead be filling up their bellies with microplastics *coated* in algae, leading to the growth of the algal blooms they normally keep in check. When these blooms die, they suck oxygen out of the water, killing off fish and other life in the area. Microplastics—with their own chemicals and agglomerated pollutants—might also directly influence the proliferation of phytoplankton.[70] Some studies have shown that particles inhibit the growth of the tiny algae while others have shown the opposite, though these experiments were done with high concentrations of microplastics in the lab, where conditions are different than out at sea. (This is a common practice in science, generally speaking: researchers do it to elicit responses from organisms, unrealistic as the concentrations may be.[71]) But with ever more petroplankton flowing into the ocean and persisting because plastic is durable by design, microplastic concentrations will be 50 times what they are now by 2100, according to one estimate.[72] "The modeling predictions are suggesting that within the next 50 to 100 years, if we carry on with business as usual, then those concentrations in the environment will reach the same kinds of concentrations as we're seeing in laboratory experiments," says University of Plymouth marine biologist Richard Thompson, who coined the term *microplastic*. "Then we'll start to see quite widespread ecological harm if we don't start to change our ways."

The ripple effects could be profound.[73] If microplastics encourage the growth of phytoplankton, they'd encourage the algae's sequestration of carbon from the atmosphere and the release of oxygen. Great for the animals on the land that breathe oxygen, but not great for fish when local water oxygen levels drop as phytoplankton die. Or instead microplastics could be discouraging the growth of phytoplankton, having the opposite effect on the cycle of carbon and oxygen in addition to shrinking a critical food source for zooplankton. (The seas have in no small part helped save humanity from itself, sequestering a third of the carbon we've pumped into the atmosphere, so these are vital processes we're talking about.[74]) It might also be that the particles are somehow decreasing the photosynthetic efficiency of the algae, as several lab studies have suggested. Basically, scientists know that phytoplankton and petroplankton are in constant contact out there, but they don't yet know *how* they're interacting. The planktonic party, though, is growing increasingly crowded with microplastics—that much is for sure.

## Fecal Matters

Under a microscope, biologists have watched zooplankton of all stripes swim through microplastic-tainted water, drawing the particles toward their mouths with the rapid beating of their little appendages.[75] The hypnotic pulsing of a jellyfish, honed by evolution to ferry food into a simple digestive system, now acts as a conveyor belt for petroplankton. Single-celled dinoflagellates take a more active approach, snagging polystyrene beads with whiplike appendages known as flagella. Wait a few hours, and you might catch the zooplankton "egesting" the particles in their fecal pellets.[76] You'd think this is where a microplastic's journey ends, but this may be its most consequential moment—planetarily speaking.

The rate at which poo sinks in the ocean is a big deal, as it turns out. Take the humble copepod. These mini-crustaceans, which sport oversized antennae that make them look like swimming T's, are everywhere in the world's oceans. Copepods eat phytoplankton and package the waste in fecal pellets that sink down the water column, where other recyclers like small fish pick at them. What's left is deposited on the seafloor as food for still more recyclers, and so the carbon first captured by phytoplankton ends up sequestered tens of thousands of feet deep. (The ugly irony is that this is how we got plastics in the first place: oil and gas deposits formed when ancient plankton died, settled to the seafloor, got buried, and transformed under tremendous pressures and temperatures into carbon-rich fossil fuels, which we turn into polymers.) But scientists are now finding microplastics in the copepods they haul out of the sea,[77] and when fed polystyrene particles in the lab the animals produce pellets that are significantly less dense and sink more than twice as slowly as non-tainted pellets.[78] If a copepod had 12,000 feet of ocean between it and the seafloor, and it dropped a microplastic-laced pellet, the feces would take 53 days longer to reach the bottom than a normal pellet. Furthermore, as the number of particles in a pellet increases, the structural integrity decreases because there's less organic material to hold the feces together. This fragmentation further slows the waste as it descends. The effect, though, will depend on the kind of plastic the copepod ingests, as denser polymers might *speed up* the pellet.[79]

Why should we care about how fast copepod crap sinks? Because the more time the pellet spends sinking—scientists call this transit the "fecal express"—the more time scavengers have to consume it. If copepods are consuming a lot of microplastic that lightens their pellets, their waste is less likely to reach the seafloor creatures that rely on it for sustenance. Conversely, if denser polymers are weighing down the pellets, the midwater creatures that rely on them for sustenance might miss out. And if creatures up and down the water column are picking at the pellet, that exposes them to the toxic particles. Another experiment hinted at the physiological consequences as well: copepods that ate microplastics consumed much less actual food, leading to energy deficiencies and stunted growth. The copepods also produced smaller eggs that were less likely to hatch, probably because they had less energy to devote to reproduction.[80]

Not content to ruin the copepod's day, the microplastic continues down the water column in the pellet, bouncing around a vast food web—eaten by one thing, pooped out, eaten again. Perhaps the particle will break free from feces 500 feet deep, or 1,000 feet, or 5,000 feet, and once again rise to the surface, only to be packaged in a new pellet and sink once more.[81] The water column, then, becomes a microplastic highway, with petroplankton in constant transit in both directions. Some get caught up in currents that move them across oceans, entering new ecosystems and growing new plastispheres to attract the hungry mouths of new zooplankton. The lucky microplastics eventually reach retirement, settling once and for all on the seafloor, maybe even in the famous Mariana Trench, to etch themselves in the fossil record.

Because petroplankton and organic plankton are both at the mercy of currents, they're amassing together in hotspots of biodiversity, just like the particles are accumulating with nutrients in sediments where life gathers. Where ocean currents meet at the surface, they form "slicks," smooth ribbons of water that belie the ecological frenzy within: swarms of fish larvae eat plankton, and seabirds and larger fish dine on the larvae. But water samples taken off the coast of Hawaii found that these slicks are also teeming with 126 times the concentration of microplastics as the surrounding waters, outnumbering fish larvae seven to one.[82] The amount of petroplankton here was up to 13 times higher than what's swirling in the Great Pacific Garbage Patch. The oceanographers further calculated that while slicks make up 8 percent of ocean surface habitat around the Hawaiian Islands, they contained 92 percent of all floating plastics, which the fish larvae are mistaking for prey. The team dissected hundreds of larval fish in the slicks and discovered particles in 9 percent of them, more than twice the rate as larvae they pulled from non-slick waters. Sure, 9 percent doesn't seem that bad, but it's actually a startling figure: these slicks are packed with larvae, so we're talking about masses of corrupted fish, which are in turn eaten by all kinds of predators.[83] Thus, as always, petroplankton climb up the food chain.

All the way up the food chain, in fact. Marine mammals feed on a variety of prey, from the littlest krill in the case of the blue whale, to squid and fish in the case of seals and dolphins. This makes the predators good "indicator" species: if a pollutant is accumulating in an ocean ecosystem, it'll accumulate in the prey these mammals are eating. Because these hunters have such long lives, biologists can see persistent pollutants build in their tissues, indicating that there's a problem in the environment at large.

But unlike with fish, biologists can't go out and catch cetaceans—whales, dolphins, and porpoises—to test them for microplastic, so they rely on necropsies of beached individuals. One team analyzed 50 such unlucky marine mammals, representing 10 species, around Britain and found that each had ingested microplastics, but surprisingly few of them.[84] Separately, another necropsy of three beaked whales that'd gotten stranded in Ireland also found low numbers of microfibers in their digestive systems.[85] It may be that these animals are good at expelling the particles, and sure enough, biologists have found loads of microplastics in walrus and seal scat in the wild.[86] Going an extra mile, in one experiment scientists fed wild-caught mackerel to gray seals at a rehabilitation sanctuary.[87] On average, they found 0.58 particles per fish, most of them being the synthetic rubber ethylene propylene, and the particles that came out of the other end of the seals were predominantly the same material. It's worth noting that an 880-pound male gray seal eats 4 to 6 percent of his body weight every day,[88] so an average of less than one microplastic per mackerel isn't much on its own, but scaled up to the diet of the seal—and any number of other piscivorous marine mammals—those particles add up.

Baleen whales, which suck in water and filter out prey, consume even more microplastics. At over 100 feet long, the blue whale snags enough krill in a mouthful to provide half a million calories—in a day, they'll eat a city bus's weight of the crustaceans, which themselves have been feeding on microplastics.[89] The second-largest cetacean, the fin whale, may be ingesting up to 77,000 petroplankton this way every day, according to one estimate, or 28 million particles a year.[90] Still another analysis of scat from baleen whales off New Zealand put the number at 3.5 million microplastics a day, or over a billion particles a year.[91] And these whales move thousands of feet up and down the water column, feeding and pooping and feeding and pooping, so they're yet another vehicle for microplastics to move between shallow and deep waters.

A bigger body means more opportunity to ingest microplastics, but marine mammals may actually have size on their side: a particle that passes through a whale or seal would cause gut blockages and other physical damage to a larval fish's insides. Experiments done on goldfish, for example, have shown that chewing on microplastics causes deep lacerations in the mouth, as well as inflammatory responses in the liver and intestines.[92] Brine shrimp die after eating microfibers, which cause severe gut damage.[93] Exposure to microplastics decreases the hunting efficiency of gobies, which also seem to confuse the particles for prey, reflecting a broader concern among scientists that fish everywhere are filling up their stomachs with microplastics, thus decreasing their appetite for actual food.[94] "This can lead to an effect on growth," says Susanne Brander, an environmental toxicologist who studies microplastics at Oregon State University.[95] "An effect on growth in an adult is an issue, but it's a much bigger issue when you're at that larval stage and you're in that rapid growth phase. And it's possible that a reduction in growth early in life could affect you long-term. Also, it could affect your chance of surviving, of being able to obtain the right food resources to grow to the point that you're able to reproduce."

Brander has demonstrated this effect, known as food dilution, in the lab. She fed high concentrations of microplastics to larval fish, but only for two hours. Brander allowed the fish to grow for two weeks, then compared them to control fish that hadn't been exposed to microplastic. "We saw that the plastic-fed fish had significantly lower mass than the controls," says Brander. "This was from a *two-hour* experiment." In the wild, fish are exposed to petroplankton constantly. And this isn't just a concern for larval fish: scientists have looked at the stomach contents of the juveniles from all seven species of sea turtle and found microplastics in every one.[96] (Adult sea turtles mistake plastic bags for jellyfish, and as the bags break down in their digestive systems, the animals excrete microplastics. The same goes for other animals that mistake macroplastics for food, like birds.) In addition to making an animal feel full, ingested microplastics may corrupt the gut microbiome, either by exposing it to an alien plastisphere or by leaching out chemicals that inhibit the growth of beneficial bacteria.[97]

Making matters worse, larval fish don't have fully developed immune systems, yet are having to fight off pathogens hitching rides in plastispheres.[98] And even before they're freely swimming animals, fish embryos floating around the ocean could soak up the chemicals that microplastics leach into the water, with as-yet-unknown effects.[99] One lab study found that high concentrations of leachate are toxic to sea urchin embryos, increasing rates of abnormal development by two-thirds.[100] Another experiment exposed the early life stages of sea urchins, jellyfish, and zebrafish to microplastics collected from beaches in the Caribbean's Guadeloupe archipelago.[101] In their time in the environment, the particles had accumulated a bevy of pollutants like arsenic, lead, and pesticides. They significantly reduced the growth of the sea urchins yet had little effect on the zebrafish, and they supercharged the pulsation of the young jellyfish, meaning the animals were expending more energy.

And keep in mind that larval sea creatures aren't just eating microplastics—they're breathing them too. Consider the lobster. As an adult, it of course lives on the seafloor, but in its larval form it gathers with other zooplankton near the surface, where microplastic counts are high. So in lab experiments, biologists exposed lobster larvae to realistic levels of particles you'd find in the Gulf of Maine.[102] Like scientists have observed with other zooplankton, the lobsters accumulated the microplastics in their guts and on their bodies, especially around the area where the youngsters' gills develop, which greatly reduced oxygen consumption. This double effect—microplastics accumulating in both the digestive system and the gills—would explain why exposure to the particles dramatically increased mortality among the test subjects.

Researchers are finding that fish and crustaceans can have a harder time clearing particles entangled in their gills than passing particles through their digestive systems. This means an animal accumulates microplastics in its gills over time, but the particles might make the journey from food to feces in a day or two, as we saw with the copepods. However, that's not to say microplastics in the gut don't cause problems in that brief residency. It might be the case that zooplankton and fishes are dealing with a kind of assembly line—or more accurately, disassembly line—of nasties: a new particle comes in, the gut absorbs chemicals and microbes, the particle goes out clean; a new particle comes in, the gut absorbs chemicals and microbes, the particle goes out clean; and on and on throughout an animal's life.

Microfibers are particularly prone to getting caught up in the frilly structures of the gills. Scientists have found that when exposed to microfibers, some species actually *increase* their oxygen consumption, and not in a good way: with particles tangled in their gills, the animals are working harder to get oxygen. That means they're expending more energy to survive. "Think about what we know about something like asbestos, and the toxicity it causes to lungs," says Brander. "It's going to be a bit different in the gills, but the same overall effect. If you have enough of the fibers adhering to the gill filaments, it's either going to cause the fish to labor more in order to breathe, or potentially to take in less oxygen overall and to lower its respiratory rate."

Either way, it's bad news for young fish and crustaceans especially, which need to devote energy to finding food so they can grow to sexual maturity as fast as possible. Predators lurk everywhere, waiting to pick off the weak—the ones not eating enough because their bellies are full of microplastic, or whose lungs are clogged, or whose immune systems couldn't fight off some microbe hitching a ride on petroplankton. Because of food dilution, even the stunted fish that *do* make it to reproductive age may find themselves at a disadvantage: the bigger a female fish is, the more eggs she makes and the greater her chances of producing offspring. "That to me is—right now, given what we know—more of a concern than the presence of a chemical on the plastic," Brander says.

Stunted fish are also less nutritious fish. If an animal has microplastic in its gills, it has to spend more energy breathing, energy it should instead be putting to growing. If it's got microplastic taking up space in its gut, it can't consume as many calories. If leachates from the microplastic are stressing its body out, it has to spend energy mounting a defense. The predator that hunts it, then, will be consuming emaciated prey, meaning it has to put more of its own energy into finding extra food. Thus the effects of microplastics could be rippling through food chains in subtle ways.

There are already signs of trouble in reef ecosystems, especially those near coastal cities. Surveys of reefs around the world, including the Great Barrier Reef, have found them swirling with microplastics from rivers, ship paint, and macroplastics that get caught up in corals and decay.[103] One study of the sediment from an Indonesian reef even found significantly higher concentrations than in bustling beaches nearby.[104]

Corals are masses of polyps, tiny animals that grow skeletons of calcium carbonate, the same material in eggshells. These polyps enjoy a symbiotic relationship with algae, which soak up sunlight to produce energy for the coral, or tentacled polyps also snag zooplankton for sustenance. Now, of course, they're also snagging microplastics. Experiments have shown that corals assume the particles into their tissues,[105] dramatically reducing growth rates of their skeletons and disrupting the relationship between polyps and algae, which reduces photosynthetic activity and therefore energy production.[106] Exposed polyps also catch way fewer zooplankton because their tentacled hands are full of microplastic, and their digestive systems grow burdened with particles. Microplastics also contribute to coral bleaching, a stress response in which the polyps release their symbiotic algae while impairing the animals' immune systems. That makes scientists extra worried that microplastics could be sneaking pathogens into the polyps, not to mention all the chemicals in the particles themselves: toxic plastic additives were found in 95 percent of coral polyp samples from the Maldivian archipelago.[107]

Corals are also incorporating microplastics into their skeletons. After exposing baby corals to particles for 18 months, in concentrations you might get in a heavily polluted coastal habitat,[108] scientists found nearly 1,400 microplastics per cubic inch of skeleton. Scaling this up to the reefs around the world, they calculated that corals sequester 44 million pounds of petroparticles each year. It's the first documentation of a living long-term sink for microplastics: like plastic is accumulating in Arctic ice and seafloor sediments, so too is it accumulating in coral skeletons, with unknown consequences for reefs. The particles might, for instance, compromise the structural integrity of corals, but no one's studied that yet.

With a multitude of species teeming in the coral reefs and planktonic crowds of the world's oceans, it's a monumental task to determine which are suffering from microplastic exposure, which are getting along fine, and which might suffer only as concentrations climb. "That's part of the struggle here, is that individual organisms could have vastly different responses to consuming microplastics," says aquatic ecologist Carolyn Foley of Purdue University.[109] Many planktonic species and corals are already stressed by warming and acidifying waters, not to mention other pollutants beyond plastic. So for them this could be a case of one plus one equaling three—or four or five or six—the stress of climate change and the stress of microplastic compounding to create an uber-affliction. Baby fish already have a hard time surviving without shelter in the open ocean. "That's a really important life stage—they have to survive through the first year," says Foley. "And they get just picked off like crazy. So if you have another stressor on top of that, that's potentially problematic."

Day by day more petroparticles flow into the sea and the burden grows. Following Jennifer Brandon's discovery of soaring numbers of microplastics in ocean sediments, another team managed to sketch the same kind of timeline, only they did it with dead fish. Natural history museums stash away the specimens that biologists bring back from the field—insects are pinned to trays, birds are taxidermied, and fish are plopped in jars of alcohol. Curators aren't typically keen on other scientists cutting open these specimens, but Loyola University ecologist Timothy Hoellein had a special case. He wanted to know if microplastics had been accumulating in fishes' guts long before anyone had spoken the word *microplastics*. So Hoellein reached out to Caleb McMahan, curator of fishes at Chicago's Field Museum, and asked if he could pretty please look inside the digestive systems of a few fishes. It just so happened that McMahan could pick out four particularly common species in his collection with specimens dating back to 1900: largemouth bass, sand shiners, channel catfish, and round gobies. Really, all Hoellein needed was their digestive systems—otherwise the fish would go back in their jars unmolested.

Hoellein found a pattern eerily similar to what Brandon had uncovered off the coast of Southern California and what others discovered in plankton samples from Scotland.[110] None of the fish specimens collected prior to 1950 contained particles. "Time was the strongest factor influencing the amount of microplastic in the fish across species," Hoellein says. There was some variability in the amount of plastic between species, which is understandable—a largemouth bass has a large mouth because it's a top predator, while the round goby is a bottom-feeder. But, Hoellein adds, "the overarching pattern was an appearance of microplastic around mid-century, and an increase thereafter." In a paper describing the findings, Hoellein drew up a graph that overlaid the microplastic counts in the fishes over the decades, Brandon's sediment counts, and the exponential increase in plastic production. The three matched perfectly.

## Plated Plastic

So microplastics are now a fundamental component of oceanic food webs. From surface waters to the deepest of deep seas, organisms consume the particles and are themselves consumed, on up the chain. And we're at the very end of that chain.

A comprehensive 2021 study gathered data on over 170,000 individual fishes, finding that overall, two-thirds of the species sampled had microplastics in them.[111] But three-quarters of *commercially fished* species tested positive for petroparticles, and that's likely an underestimate, the authors noted. They added that the occurrence of plastic in fish doubled between 2010 and 2019, and that fish in the heavily polluted waters of East and Southeast Asia tallied the highest levels of microplastics. These are species that feed some 2 billion people.

The critical consideration here is a process called translocation. When a fish eats a microplastic, the foreign object may well bounce around the digestive system and come out the other end intact. But the concern, especially with the smallest of particles, is that they pass through the gut wall and into the blood and other tissues. If salmon consume microplastics but the particles just hang around the gut, that's no big deal for fans of salmon—the digestive bits are chucked out, and we only eat the muscle and skin and roe. But if the particles are translocating through the gut and into muscle, then that's an avenue for oceanic microplastics to get into our own stomachs.

Scientists have indeed found microplastics in the livers and muscle tissue of fish caught in the wild,[112] and in the lab, they've fed microplastics to fish and watched the particles translocate to organs and muscles.[113] One group of researchers collected fresh oysters, prawns, squid, crabs, and sardines from a market and tested their edible bits, finding particles in each.[114] The sardines were far more contaminated than the other seafoods—eat three of the little things and you'd also get a grain of rice worth of microplastic. But a 2021 review of 132 papers on translocation in fish found limited evidence that microplastics are contaminating the meat we humans eat.[115] So the usual caveats apply here: samples could get contaminated between the sea and the lab, these groups are using their own testing techniques, and the littlest of particles are evading detection.

Theoretically, nanoplastics should more easily work their way through a fish's digestive system and into its bloodstream, which would carry the particles to all the organs, including the brain. And indeed, scientists have watched this happen. A team laced algae with 53- and 180-nanometer polystyrene particles, then fed it to freshwater crustaceans, then fed those to freshwater carp. Compared to a control group, the carp that fed on crustaceans tainted with the 180-nanometer particles were much more active and ate much faster. The carp exposed to the 53-nanometer particles ate slower and had to swim longer distances to feed, which means the fish had to spend more energy to hunt while also exposing themselves to more predation out in the open. Sure enough, when the researchers opened up the exposed carps' brains, they found nanoplastics in them all. The exposed fish had also lost weight and had less water in their brains.[116]

The experiment documents trophic transfer in action, as the particles migrated from vegetation to herbivore to predator, and confirms that nanoplastics can translocate from the gut of a fish into its blood, then into its brain. That would mean the particles evaded the blood-brain barrier, a sort of force field that keeps out nasties like bacteria and viruses, and the dramatic behavioral change in the carp suggests brain damage. Yes, the experiment exposed the fish to higher levels of nanoplastics than they would find in nature, but it also only ran for two months, while carp can live for 10 years. That is, the fish aren't encountering this much nanoplastic day to day, but year to year, they may be accumulating the stuff in their brains—remember that a load of laundry releases hundreds of trillions of nanoplastics, so the oceans are growing increasingly burdened with the particles. Also, the researchers only noticed the weird behavior halfway through the experiment, suggesting that *accumulated* particles, not an initial dose, were the problem. (Scientists have also been finding worrying behavioral changes in animals exposed to even small concentrations of microplastics for short periods of time: hermit crabs kept for five days in water containing just 25 particles per liter took far longer to evict rivals from the snail shells they need for protection.[117]) And these scientists were experimenting with virgin plastics, whereas the ones journeying around the oceans are magnets for pollutants and microbes, which may exacerbate the effects on the brain. Because they're so tiny, nanoplastics have large surface areas relative to volume—remember our potatoes—meaning that if the particles are infiltrating fish brains out in the wild, they're delivering lots of whatever they've accumulated from the environment.

In a similar experiment, scientists fed nanoplastic-tainted algae to the same freshwater crustaceans, then fed those to Chinese rice fish, and finally fed those to dark chub, a predatory fish.[118] They noticed the nanoparticles damaging the intestinal walls of the crustaceans and confirmed that the plastics jumped to the rice fish and finally the dark chub—trophic transfer, through and through. In addition, the researchers exposed each species to nanoplastic water and found similar behavioral changes as in the carp study: the rice fish and dark chub both explored significantly less area in their aquariums. They also soaked the fertilized eggs of the rice fish in the nanoplastic water for 24 hours, then watched as the embryos developed. The particles penetrated the walls of the embryos, and the juveniles born from these were also contaminated with particles, showing that nanoplastics—in high concentrations, at least—corrupt fish before they're even born.

Unfortunately, biological processes turn microplastics, which might not pass through barriers in the body, into nanoplastics, which do. In one study, researchers fed microbeads to krill, then dissected the animals.[119] Going in, the particles were 31 micrometers in diameter, but in the gut they disintegrated into fragments that were on average 7 micrometers, with some less than 1 micrometer. So a microplastic in a krill gut might be too big to pass into other tissues, but it breaks into smaller particles that could migrate through the body. On a larger scale, that means zooplankton like krill may work as innumerable little factories, turning microplastics into nanoplastics they poop into the world's oceans to be eaten and breathed by other creatures. It's not just sunlight and friction, then, that break apart plastics, but animals as well.

The accumulation of microplastics and nanoplastics in fish is especially relevant for species farmed in coastal enclosures: aquaculture now makes up half of global fish production.[120] The feed that fuels this massive industry is made of either small, ground-up whole fish like anchovies and sardines, or fish byproducts—the guts and such that humans don't typically eat. Some 11 billion pounds of this fishmeal are produced annually.[121] We know for certain that fish retain microplastics in their guts and that, according to one study, commercial feeds average 56 particles per pound.[122] Biologists have further confirmed that farmed fish accumulate microplastics in their gills and guts that match what was in their feed—the higher the plastic contamination in the fishmeal, the higher the contamination in the fish.[123] The particles that fish don't retain in their bodies, they excrete to drift down through the enclosure and enter the aquatic environment. Plus these fish are all penned in with plastic netting that's also shedding microplastics.

Because we don't generally consume fish guts, our exposure here may be low. But particularly small fishes, like sardines, are eaten guts and all, and so are bivalves like clams, mussels, and oysters. These filter-feeders suck in water and strain out the organic material—but also now microplastics: scientists consistently find hundreds of particles in a single animal.[124] Matthew Cole, at the Plymouth Marine Laboratory in the UK, has shown that 300 mussels can remove a quarter of a million particles from the water every hour, which they pass in their feces. Cole's actually playing around with a way to encourage the growth of mussels in places particularly polluted with microplastics, like harbors or the mouths of rivers or where treatment plants release their effluent, and engineer a system that would gather their feces for removal. "In an ideal world, we wouldn't even have to be thinking of using mussels in this way," Cole says. "But unfortunately, plastic is so prolific, and it has so many entryways into marine ecosystems, that it's really hard to stop all of the leakage points."

If bivalves just pooped out all the microplastics, we'd only be eating the ones in their guts when they died. But one team of scientists even cleared the digestive tracts of mussels and oysters before dissolving the specimens in acid and still found microplastics, suggesting translocation from the gut to other tissues.[125] (Microplastics are also in scallop meat, even though we only eat the animal's muscle.[126]) Based on this tally, the researchers calculated that a European consumer eats up to 11,000 microplastics a year from mussels and oysters alone.

And so the unchecked torrent of microplastics into the seas comes back to bite us—or more accurately, we're now biting the microplastics. From the tiniest of plankton to the biggest of whales, petroparticles have thoroughly infiltrated the food chain. The good news is that wastewater treatment has kept loads of microfibers from flushing into the environment. The bad news is that when a facility captures these particles, it does so in the human waste that farmers spread on their fields as fertilizer. As microplastics swirl in the oceans, so too have they fouled the land.

CHAPTER 3

# A Land Corrupted

Everywhere else in the world rain brings life, but in Washington it also brings death. Back in the 1980s, locals began noticing mass die-offs of coho salmon in the state's streams—"unexplained acute mortality," in the scientific parlance. Year after year, to this day, salmon turn up dead after heavy rains. And *only* coho salmon, as other fish in the same streams, even the related chum salmon, make it through fine.

A team of scientists at the University of Washington had a hunch that the culprit wasn't the rain itself, like an acid rain kind of situation, but stormwater flowing into streams from human populations. Indeed, the more heavily trafficked roads near a stream, the higher the mortality rate. Which meant the researchers were looking at a long list of suspects embedded within a long list of suspects: motor oil, antifreeze, tire particles, even windshield wiper fluid could be at fault, but each potential killer itself contains a cocktail of chemicals. And only the manufacturer knows how to make that cocktail. "If you look at a tire, it doesn't come with an ingredient list," says Edward Kolodziej, an environmental engineer and chemist at the University of Washington. "There's no ingredient list on your shoes, no ingredient lists for your motor oil or your antifreeze. And it's not just that there's no ingredients—it's all protected as confidential, proprietary business information. That's just a huge barrier to understanding what's out in the environment, and how it might be harmful for either organisms or people."

And so Kolodziej and his colleagues had to find the coho killer the hard way. From stream samples, they detected a chemical signature from tires, so they exposed juvenile salmon to the leachate from these particles. Sure enough, it killed the coho, but not chum salmon. To narrow down their suspects further, they created different leachates with certain tire chemicals removed, then exposed the salmon again to each. Throwing out suspects one by one over the course of two and a half years, eventually the detectives zeroed in on N-(1,3-dimethylbutyl)-N′-phenyl-p-phenylenediamine, known more mercifully as 6PPD, which on its own killed the coho.[1]

In the early days of automobiles, tires were made from natural rubber. Slice through the rubber tree's bark and it'll spit out latex, a white ooze that is itself a natural polymer. From this latex you can extract rubber, a stretchy yet tough material, properties that made it a coveted ingredient in all manner of products in the early 20th century, from boots to electrical insulators to car tires. But as demand soared during World War II, the supply of natural rubber from Southeast Asia was cut off, forcing chemists to concoct alternative polymers derived from fossil fuels.[2] Today, tires are made of both natural and synthetic rubbers, in addition to a bevy of fillers, pigments, and antioxidants, plus you've got steel wires and nylon and polyester fibers thrown in for support.[3] So a tire is a plastic, and a particularly complicated one at that.[4]

Manufacturers had at first coated their tires in wax to ward off ozone, a persistent ground-level pollutant. "If you've ever seen an old tire, and it has just these huge deep cracks in it, that's what ozone does," says Kolodziej. Looking for a more efficient alternative to wax, in the 1950s the US Army developed 6PPD, a chemical that reacts with ozone faster than the gas can react with the tire itself.[5] "6PPD was also good because it's actually insoluble in rubber," says Kolodziej. "Over timescales of years, it literally tries to leave the rubber by diffusing out to the surface. So as it reacts with the ozone, it's kind of replenishing itself to make like a protective film barrier to protect the tire." This chemical process turns 6PPD into 6PPD-quinone, and it's the kind of transformation Imari Walker warned us about: it isn't just the ingredients in plastic products that scientists have to investigate, but how those ingredients transform once in the environment.

Really, you couldn't invent a better way of generating microplastics than car tires, the polymers subjected to the twin stresses of heat and friction. As the tire rolls along, completing 750 rotations every mile, the road abrades it, sheering off particles.[6] This friction generates heat—a half hour of highway driving raises the tire's temperature by 50 degrees Fahrenheit—which only assaults the polymers further. If you live where roads are particularly crummy and the temperatures particularly high, your tires emit particularly high amounts of microplastics.

All told, tire microplastic emissions per capita range from half a pound to 10 pounds a year, depending on your country, the global average being almost 2 pounds.[7] Even in a small nation like Sweden, total vehicle tire emissions could be up to 29 million pounds a year. In China, it'd be more like 1.6 billion pounds, and in the car-crazed United States, as much as 3.3 billion. One estimate puts annual global tire emissions at 13 billion pounds, enough material to fill 31 of the biggest container ships, each of which is a quarter mile long.

But tire microplastics aren't ending up in container ships—they're ending up in surface waters like the streams of Washington. "I do think especially in freshwater habitats, tires are probably one of the most ubiquitous microplastics, and maybe the one we should be paying attention to," Kolodziej says. But Kolodziej and his colleagues can't yet say why 6PPD-quinone is so toxic to coho salmon, and why it's not toxic to chum salmon.[8] In a follow-up study, Canadian scientists found that the substance also kills rainbow trout and brook trout—in realistic concentrations, mind you—but not white sturgeon or Arctic char, even in super high doses.[9] So 6PPD-quinone could be dispatching still more species out there without humans knowing about it. "It is known in toxicology that chemicals can have drastically different effects in different species," Kolodziej says. Also, 6PPD-quinone may harm other species in ways that aren't as obvious as mass mortality events—for instance, birth defects and reproductive complications. "You don't have to kill something to harm something."

Take the caddisfly. As a young wormlike larva living on a riverbed, it stitches together bits of sediment with silk to form a protective cone. But biologists are finding that caddisflies now also incorporate microplastics into their cases, and in fact the larvae prefer the fragments over sediment in the early stages of construction, since plastic is lighter than minerals.[10] The more microplastics in there, the more the structure weakens. So even if the particles don't kill caddisfly larvae by way of ingestion, they could indirectly kill the animals by compromising their defenses against predators. And because a normal caddisfly cone contains sediment, it helps the larva blend into the riverbed, but if the cone is instead made of multicolored microplastics, that'll make it stick out. The effect of microplastics here is subtle, but the consequences could ripple across the ecosystem, as these critters eat aquatic vegetation to keep the river from getting overgrown.

Stormwater runoff is dosing these river ecosystems with microplastics on a massive scale.[11] Researchers working in northwest England found up to 50,000 particles per square foot of river sediment,[12] and Amsterdam's famous urban canals are similarly polluted.[13] Two rivers in Los Angeles send over 2 billion bits of plastic to the sea every three days.[14] Another survey of the Ganges put the figure at up to 3 billion particles released daily into the Bay of Bengal.[15] "It shows us how rivers can be a huge conveyor belt for plastics going out to sea," says University of Plymouth microplastics scientist Imogen Napper, who led the Ganges research. So we need washing machine filters and better clothes to keep microfibers from leaving the home in the first place, because once they reach a river, it's too late to keep them from reaching the sea.

We can, though, snag larger plastics as they float downstream—or I should say a family of barges in Baltimore can: Mr. Trash Wheel, Professor Trash Wheel, Captain Trash Wheel, and Gwynnda the Good Wheel of the West.[16] These are very much their real names, and they're very much effective at gathering plastic. Containment booms extend out into the river, down to two feet deep, funneling trash into a conveyor belt—powered by a waterwheel and topped with big anime-style eyes—that dumps the refuse into a bin for disposal. (According to the profiles on their website, the barges' likes include pizza boxes and birdwatching, while Mr. Trash Wheel dislikes ducks "but denies ever eating one." Gwynnda isn't a fan of the Salem witch trials, naturally.) Since 2014, the trash wheel family has collected 3.5 million pounds of debris in Baltimore, including 830,000 plastic bags and 1.5 million plastic bottles. (And one each of—and I'm not making this up—a keg, a guitar, and a ball python.) If we can get these barges in highly populated rivers around the world, it'll go a long way in keeping macroplastics out of the sea. The challenge, though, is traffic: you don't want boats steaming through your containment booms. And where there's the most traffic, there's the most people and the most refuse. Still, trash wheels working along shorelines can at least turn down the faucet of pollution.

But back to the microplastics. In 2019 scientists reported finding high concentrations of black particles in sediments around the San Francisco Bay, which they confirmed to be bits of tire.[17] They also found astronomical levels of microfibers, which entered the bay both in effluent from wastewater treatment plants and just from the street-level bustle of humanity: our synthetic clothing and shoes shed particles as we walk.

Humans dust the ground with microfibers everywhere they go, including Mount Everest, where a liter of snow contains up to 120 particles shed from hikers' performance clothing.[18] Not to make you feel bad about your outdoor activities, but you never leave a place pristine, no matter how carefully you clean up after yourself.

Also, cigarette butts—they're plastic too. Each is made of 15,000 cellulose acetate microfibers loaded with chemical compounds, more than 40 of which are known to be toxic to aquatic organisms.[19] (John Wesley Hyatt's synthetic billiard balls were made of highly flammable cellulose *nitrate*—it goes without saying that you wouldn't want that in a cigarette.) Of the 6 trillion cigarettes smoked worldwide each year, 4.5 trillion become litter, contributing 660 million pounds of microfibers to water bodies, and when smokers step on their cigarette butts, they prime the things to disintegrate.[20] Cigarette butts are consistently the most-collected debris on beach cleanup walks.[21] Think about how terrible smoking is for you, then think about how cigarette microfibers diffuse that terribleness into the environment. "They're a double-edged sword, because it's not just the physical plastic particles, it's also the fact that you've got hundreds of different compounds after smoking," says ecologist Dannielle Green, who studies the pollutant at Anglia Ruskin University. "So they're like their own kind of toxic bombs, potentially, as litter. And it's just this crazy conception that it's still relatively socially acceptable. You see it every day, people just flicking these things."

So when it rains on the San Francisco megalopolis, stormwater flushes tire particles and clothing microfibers and cigarette butts into the bay, to the tune of 7 trillion microplastics each year, according to one estimate. (I am therefore skeptical of schemes in which engineers mix plastic waste with other paving materials to make plastic roads.[22] Yes, that keeps plastic bottles and such out of the environment, but as the roads deteriorate, the debris end up in the environment anyway as microplastics. Plastics are already on roads, in fact, in the form of markings that abrade in traffic, hence needing to be constantly repainted.[23]) A separate study in 2021 found that 25 particles per liter of stormwater run off into watersheds around the San Francisco Bay, mostly fibers and rubbery fragments.[24] These concentrations are way higher than what you'd find coming out of a wastewater treatment plant.

Tire particles aren't just spewing from roads, but from . . . soccer fields.[25] To give artificial turf the springiness of real grass, manufacturers add ground-up car tires, known as rubber crumb. A field uses up to 260,000 pounds of the stuff.[26] As players stomp the turf, flinging particles into the air, the pitch loses so much of this microplastic that periodically someone has to come and top it up with more rubber crumb. A single field sheds thousands of pounds of the stuff a year, a black grime that creeps away from the field and into stormwater drains. Estimates vary widely, but in Europe alone, over 50,000 artificial pitches could be emitting 160 million pounds of rubber crumb into the environment each year.[27] (Less well studied is how regular artificial lawns without rubber crumb—one you'd buy for your yard—shed microplastics, given the blades of grass are made of plastic.[28])

Like microplastics in general, there's no removing tire particles now that they've saturated the environment. We might build better catchment systems to protect coho salmon, sure, but let's be realistic here: each car tire is a proprietary amalgamation of chemicals, any one of which could already be imperiling any number of species. "Generally, we figure out that chemicals are harmful after the fact, after they've become systematic pollutants, where they're all over the place," says Kolodziej. "And then you have this big widespread pollution issue that typically is pretty dang costly to fix."

But there are ways to mitigate the tire problem: bolster public transportation, which means fewer cars on the road flinging tire particles, and deploy the humble rain garden, really just a depression filled with dirt and plants instead of pavement. True to its name, it gathers rainwater from roadsides, allowing the fluid to soak into the ground instead of flooding the surrounding area. One study of a rain garden in the San Francisco Bay Area found that the green space retained up to 95 percent of microparticles that flowed into it.[29] At the same time, a rain garden beautifies the urban landscape. "There are so many multiple benefits of green infrastructure—I'm a huge fan," says Alicia Gilbreath, an environmental scientist at the San Francisco Estuary Institute, who led the study. "I think one challenge is it's expensive, and we can't put it everywhere." Not to mention that someone has to maintain the sites as they grow polluted over time.[30]

That and urban microplastics aren't all flushing out to bodies of water, but they are also corrupting parks and other green spaces within cities. You might think of urban trees as the hardiest of hardy organisms, able to withstand a constant bombardment of pollution like car exhaust. But scientists are concerned that they too are getting dosed with road microplastics, with as-yet-unknown consequences for their health. "The stormwater is conveyed into a green space, and you have these microplastics that are essentially seeping down through these soils," says environmental scientist Timnit Kefela, who's studying microplastics at the University of California, Santa Barbara. "It's a suite of contaminants entering these systems, and we're not paying attention to how they could be potentially affecting them. Are we having a decrease in this soil microbial population? Are you changing the microbial dynamics in any way that could in turn be very harmful to the plants?"

You might be wondering what the big deal is, since we don't rely on urban green spaces for food or anything. Urban folk do, however, rely on them to make life bearable. The built environment—roads, sidewalks, structures—absorbs the sun's energy during the day and slowly releases it at night. Consequently, urban areas can be 20 degrees Fahrenheit hotter in the summer than surrounding rural areas, where vegetation cools the landscape.[31] So as the planet warms, healthy green spaces will be ever more critical for attenuating this urban heat island effect, as it's known. All the while, humans are migrating to cities en masse: right now half of the world's population lives in urban areas, but that will jump to 68 percent by 2050, according to the UN.[32] And as urban populations grow, so too must urban trees. Rain gardens would pull double duty here, both lowering temperatures and snagging microplastics before they flow out to sea. "I do think that paying attention to these urban environments," says Kefela, "especially these urban green spaces, gives us the opportunity to ensure that we have a much more inclusive environmental future that is healthy for everybody."

## Fields of Fibers

We'd hope that our crops—far away from metropolises brimming with microplastics—would escape contamination. But in actual fact, they're getting it the worst. In an ideal world, treatment facilities would safely dispose of the microfibers and microbeads they filter out of wastewater. Instead, they accidentally sequester the particles in a byproduct known as sludge: human waste and other organic matter that's sold to farmers as fertilizer. The US generates over 28 billion pounds of these biosolids a year, half of which is applied to soils—the rest is landfilled or incinerated.[33] According to one estimate, in North America nearly 700 million pounds of microplastic are spread on fields each year via sludge, while in Europe it's closer to a billion pounds.[34] "We are removing plastic from wastewater, just to redistribute it in a diffuse way over the landscape," says Luca Nizzetto, a senior research scientist at the Norwegian Institute for Water Research.[35] "Actually, in places where we *produce food*." (Which, by the way, means that wealthier urban populations are literally shoveling their plastic pollution into poorer rural areas.)

To determine how bad the problem has gotten, Nizzetto took sludge samples from eight of Norway's wastewater treatment plants and tested them for microplastics.[36] (Which involved picking through waste, by the way. "Very messy business," Nizzetto says. "But someone had to do that.") The samples averaged over 1,000 particles per liter of sludge, the highest count being 3,400. Around 38 percent of the particles were beads, 32 percent were fragments, and 29 percent fibers, the rest being glitter. (Yes, glitter is microplastic, as if we needed another reason to never use it ever again.) Combining this data with rates of sludge use, Nizzetto estimated that each year 584 billion microplastic particles are applied to soils in Norway, a country of just 5.4 million people.

Norway's neighbors aren't faring much better. The sludge from seven facilities in Ireland—home to about the same number of people—tested for an almost identical amount of microplastic.[37] A report commissioned by the Swedish Environmental Protection Agency found that every hour, over 3 million particles entered one of the country's wastewater facilities, 99 percent of which was sequestered in sludge, and this was a plant that served 14,000 people.[38] In Germany, a survey found quadruple the number of microplastics in sludge that Nizzetto did,[39] but another German study reported tens of thousands of times more, because it counted particles as small as half a micrometer.[40] Again, just because scientists can't test for it doesn't mean it isn't there—any analysis of sludge is underreporting the true number of particles, especially considering that trillions of nanoplastics release from a load of laundry.

No matter its country of origin, sludge is packed with microfibers, microbeads, and other forms of microplastic. If you'll remember the study that calculated that we've washed over 6 billion pounds of microfibers into water bodies since 1950, those researchers also estimated that the land has received 4 billion pounds of microplastics via sludge. (Scientists in California provided some useful perspective here by estimating how much microplastic the state expels into the environment each year, in terms of toys: up to 9 million pounds, or 80 million rubber duckies.[41] If humanity has expelled 10 billion pounds of just microfibers—both onto fields and into bodies of water—that'd work out to 90 billion rubber duckies.) If all those particles are compromising the health of agricultural soils, "then we can expect cascading ecological implications," says Nizzetto. "Maybe in some cases, in some heavily impacted soil, we are already facing the situation where this type of impact may occur."

At the risk of belaboring the point, plastics are enduring by design. In New York, researchers discovered that microfibers had persisted for at least 15 years in treated soils.[42] Chile has kept meticulous records of sludge use in its fields, as producers have to report each application to the government, which allowed scientists to single out a region in which 30 abutting fields had between one and five sludge applications over a decade.[43] They found that with each treatment, microplastic concentrations in the soils climbed, showing that successive applications only pile on the contamination.

What's not clear is how the atmosphere might also play a role here. A desiccated field will be scoured by winds, which carry dirt particles far and wide—remember the Dust Bowl. It's safe to assume that microplastics would go along for the ride, synthetic materials being less dense than natural ones. Also not clear yet is the role of agricultural runoff. In Canada, Nizzetto and his colleagues found that microplastics accumulated in the upper layers of soil following a sludge application, then declined over time as the particles migrated to deeper layers.[44] However, another field saw this same loss of microplastics in the upper layers of soil without migration to deeper layers, which may be due to heavy rains washing away the particles. That'd mean fields switch from being accumulators of microplastics to sources of microplastics, distributing the particles to nearby water bodies. Flooding could make matters even worse: a deluge can deposit up to 40 times the normal amount of macroplastic waste into rivers, which may hold true for microplastics as well.[45]

Even if a farmer isn't keen on spreading human waste on their fields, the alternative isn't plastic-free either. Organic fertilizer is made of biowaste from households, restaurants, and food manufacturers—that's where your apple cores go after you chuck them in the compost bin. But all that food is often gathered in plastic bags, *then* chucked into the bin. A composting plant removes as many of these as it can, but some still get chewed up and incorporated into fertilizer. German researchers tested samples from such biowaste treatment plants and found that in their country alone, farmers apply over 2 trillion microplastics to fields each year, and that's only counting particles bigger than a millimeter.[46]

In applying sludge and biowaste, farmers may also be applying pathogens. Scientists have found bacterial communities on sludge microplastics to be way more diverse than particles plucked from raw sewage and the effluent that flows out of a wastewater facility.[47] They've also discovered a number of bacteria that cause human intestinal infections on microplastics downstream of treatment plants, so the microbes are using the particles as rafts.[48] Another experiment found that when applied to sludge in the lab, microplastics grew far more biofilm than sand particles, in turn accumulating up to 30 times the abundance of genes associated with antibiotic-resistant bacteria.[49] These researchers reckoned the issue was an especially populous bacteria on the particles called *Novosphingobium pokkalii*, which secretes a glue-like substance that other bacteria stick to—the microplastic particle, then, snowballs through the sludge collecting microbes, far more efficiently than natural particles like sand.[50] Extra worrisome is that bacteria pull off a trick called horizontal gene transfer, in which DNA swaps between individuals instead of from parent to offspring, and that can even happen between different species of bacteria. In this way, the genes that code for antibiotic resistance might spread rapidly through the crowded plastisphere, and as the particle rafts across bodies of water it might circulate these genes among microbes in new ecosystems.[51]

You might say, Well, yeah, sludge is human waste—did we expect the microplastics to come out sparkling clean? Not in the least bit, but we do have different expectations for how those particles might behave after they've been applied to a field. Given their low density and galaxy of shapes, microplastics move about the environment differently than organic sludge particles. They may be more prone to getting caught up in the wind, for instance, or more easily flow off fields in runoff. And what happens when a drought strikes and sludgy fields desiccate? What role might microplastics play in the ensuing dust storms as vectors of pathogens and antibiotic-resistant bacteria?

And sludge and biowaste aren't the only way microplastics are getting into fields. A farmer might apply coated fertilizers, plant food packaged in plastic capsules, also known as prills.[52] This microcapsule slowly releases nutrients for a more uniform delivery to the plants, plus the farmer saves money on the labor required for repeated fertilizer treatments. But because the spheres are so small—about the size of nurdles—there's no recovering them, so they're left to break down in the soil.[53]

Or, in Japan's Ishikawa Prefecture, to wash off of fields and into the sea. Where the Tedori River meets the Pacific Ocean, it splays out in a great "alluvial fan," where paddy fields grow. During the irrigation season, farmers flood their fields for the young rice, turning the landscape into a stained glass of plots glinting in the sun. Farmers here have used coated fertilizer for decades, and intensively so: between 1976 and 2016, Japan imported and manufactured almost 5 billion pounds of the stuff.[54] "This is a great advantage for Japan, where the population of farmers is decreasing and the farming population is aging," says Naoya Katsumi, an environmental scientist at Ishikawa Prefectural University.

The disadvantage is that Katsumi finds copious prills in Ishikawa Prefecture's paddy fields and surrounding environment. "They are not collected, but accumulate in the soil, or some of them are transported to the ocean," Katsumi says. In the same way that soils treated with sludge accumulate microfibers and microbeads, so too are these rice fields amassing microcapsules, which shrivel up like tiny empty turtle shells as they expend their payloads. Katsumi even found one field that had never been treated but still had microcapsules in its soil, likely because farmers share water for irrigation. This water eventually flows out to sea, carrying huge quantities of plastic. Sampling eight sites on the shoreline of the Tedori River's alluvial fan, Katsumi charted how microcapsule deposition varies throughout year: when farmers aren't irrigating, up to 140 prills still turn up per square foot of beach, while in the irrigation season the number soars to 7,000.[55] Katsumi reckons that paddy fields across Japan have accumulated a billion pounds of microcapsules—a compounding environmental price for agricultural progress.

## The Rise of Plasticulture

Linguistically speaking, the only thing more peculiar than Emery Emmert's own name is the name of the revolution he conjured out of the dirt. A horticulturist at the University of Kentucky, Emmert was on a mission to feed the world, and he certainly lived in the right time for that.[56] It was the expansionist 1950s, and the polymers that had been developed so frenetically during the war were infiltrating every aspect of human life—Tupperware, plastic furniture, the Hula-Hoop. Emmert, though, saw in the miracle material a magic wand to wave over crops, growing them faster and bigger: plasticulture, as it would soon be known, without a hint of apprehension.[57]

Emmert invented the minimalistic "field greenhouse," a frame of lightweight wood wrapped in plastic sheeting.[58] Because it protected crops from pests, extended a growing season by keeping out the cold, and increased yields, the simple technology quickly went global. For European and Asian countries recovering from the devastation of the war, the field greenhouse was cheap and effective, feeding more mouths with the same old crops. Emmert also developed a simplified housing called a row cover, plastic arched over the length of a row to create a hangar with plants instead of planes inside. This, too, protected crops from pests and the elements, only on a much larger scale than a field greenhouse. The calculus for a farmer here was easy: invest in some cheap plastic, and you'll probably end up with more food to sell.

Emmert further simplified plasticulture with plastic mulch. Whereas the field greenhouse and row cover act like shelters for crops, this polymer sheet is more like a sleeping bag for the plants. Or, more accurately, for the dirt: a machine stretches the plastic taut over the mound, all the way down the row, then the farmer pokes holes in the sheet and deposits the seeds.[59] Like the two other revolutionary tools of plasticulture, plastic mulch increases yields,[60] but the mulch is unique in that it locks nutrients into the soil, improving the plants' overall health. Depending on the crop, you might deploy white mulch to reflect the sun's energy, cooling the soil, while black has the opposite effect. Clear mulches don't really absorb solar radiation themselves, but they still trap moisture and heat between the film and the soil.[61] So much heat, in fact, that farmers use clear mulch to bake any insects and weeds trapped inside, a technique known as soil solarization.[62]

You can see where this is going. Other than a car tire, there's no better way to abuse plastic. Farmers are stretching it taut, providing mechanical stress, and leaving it out in a field for UV bombardment to break apart polymer chains. The film is notoriously difficult—impossible, really—to remove from fields in one piece.[63] Plastic mulch eviscerates for a number of reasons beyond application and removal, like scouring by strong winds and pelting by downpours, leaving tatters of plastic in the field. Even farm animals chew the stuff up: one study found that when sheep are let loose to graze on the remnants of crops grown with plastic mulch, they excrete nearly 500 microplastics per pound of feces.[64] Harvested fields end up covered in so much debris, it looks like a plastic bomb went off. The mulch that farmers *do* manage to pull off their fields can't be recycled because it's caked in dirt and fertilizers and pesticides.

So instead they'll often pile up the mulch and burn it, dosing not only their own soils but surrounding lands and the atmosphere with petrochemicals and petroparticles.

In 1982, China used 13 million pounds of plastic mulch to cover around 460 square miles of farmland.[65] By 2011, the country was using 2.6 billion pounds over 77,000 square miles. That's enough plastic mulch to cover Nebraska. Some 23 billion pounds of the stuff is now used worldwide.[66] Depending on factors like the kind of polymer and the weather, the top foot of a farm's soil may contain hundreds of pounds of residual plastic per acre.[67] And the longer a field has been treated with mulch, the more particles accumulate, like microplastics from successive sludge treatments.[68]

As plastic mulch works deeper into the soil, it's no longer exposed to the main stressor breaking it apart: sunlight. It might take three centuries for a thin sheet of plastic to disappear from a field—well, not so much disappear, but break into microplastics we can no longer see.[69] So while plastic mulch does increase yields in the short term, in the long term it's a nasty bargain for farmers. One study in southwestern China found up to 18,000 microplastics per pound of soil,[70] and another estimated that over a billion pounds of plastic mulch has accumulated in fields across the country, reducing cotton yields by 6 to 10 percent. "Immediate actions should be taken," the researchers write, "to ensure the recovery of plastic film mulch and limit further increase in film residue loading to maintain the sustainability of these croplands."[71]

Like prills, plasticulture is loading the oceans with microplastics. Sprawling 122 square miles on the southern coast of Spain is a nearly unbroken mass of greenhouses known as the Mar de Plástico—the Sea of Plastic.[72] Each year the region produces nearly 7 billion pounds of fruits and vegetables for export around Europe, as well as an astonishing amount of microplastic for export to the Mediterranean Sea. As expected, the plastic films that cover these greenhouses break apart when bombarded by winds and UV radiation. Taking sediment samples from seagrass beds along the coast, researchers could actually chart the rise of greenhouse farming in the Sea of Plastic.[73] Starting in the mid-1970s, microplastic concentrations climbed rapidly, and now up to 800 particles settle on a square foot of seabed each year. Plot the microplastic counts over time and they match the growth in greenhouse surface area in the Sea of Plastic over the last 50 years, just like Jennifer Brandon's sediment samples and Timothy Hoellein's fishes expose the growth of plastic production generally.

Plasticulture, prills, and mountains of sludge are conspiring to poison the world's soils and seas with ever more microplastic. Particle by particle, the stuff is accumulating and persisting, and scientists are finding troubling signals that microplastics are bad news for crops. "The urgency for coordinated and decisive action cannot be understated," wrote the Food and Agriculture Organization of the United Nations—a group not exactly known for histrionics—in a major 2021 report on plastics.[74] "Measures to both reduce the direct environmental harm caused by agricultural plastic pollution, and the indirect impacts of [greenhouse gas] emissions associated with the use of petroleum-derived plastics, need to be implemented as a matter of priority."

## Digging Up Dirt

Like marine mammals serve as indicator species in the sea, the humble earthworm serves as a wiggling barometer of soil health. Dirt is both their home and their food—they eat organic matter and leave their waste as fertilizer, called castings. Their movement also aerates the soil, further boosting its productivity. A happy earthworm population, then, makes for a happy soil ecosystem. But as petroparticles have spread across the earth, scientists worry that the pollution could be triggering a cascade of effects through ecosystems underfoot, and they're looking to earthworms for early warning signs.

We already know that earthworms transport microplastics up and down and all around soil. In one experiment, researchers gathered dirt from a meadow in Berlin, brought it into the lab, and sprinkled leaf litter on the surface for food, plus spherical white microplastics.[75] Then they added a common kind of earthworm and waited as the subjects milled about, feeding on the leaf litter and depositing their casts to form the walls of their burrows. After 21 days, the earthworms had managed to redistribute particles throughout their tunnels, and the researchers also found beads in casts and sticking to the earthworms themselves. In a similar experiment, another team used the same species of earthworm but spiked their food with different concentrations of microplastic.[76] They used terrariums, like a more professional kind of ant farm, so they could observe the worms' burrows in cross-section. After letting the earthworms roam for two weeks, they froze the mini-habitats—worms, dirt, and microplastic all. This gave them a complete snapshot of each burrow, which showed that the worms "significantly enriched" the walls of their homes with particles, especially at higher concentrations. The researchers also weighed each earthworm and found that the subjects exposed to microplastic at any concentration lost far more weight than worms with no exposure, which jives with their previous experiments that found microplastics reduce growth rate and increase mortality in the animals.[77] And we know that earthworms in the wild are consuming microplastic, as scientists have turned up particles in casts collected from the field. Furthermore, earthworms seem to process microplastics into smaller bits, like krill do, meaning they're both spreading the pollutant through the soil and acting as biological factories to churn out smaller, more numerous particles.

Ants, termites, and even mammals, like gophers and groundhogs, are also mixing microplastics deeper into the ground. Lizards, too, and even birds, like the burrowing owl, track petroparticles into their caves, like workers track nurdles out of plastics factories in their boots. Decaying plant roots also leave cavities, known as biopores, into which particles tumble, or a rain shower gathers microplastics at the surface and drags them deeper as the water trickles through the dirt. Soil also develops fissures when it dries out, allowing particles to fall in, then smooths over once more when rains return, locking the microplastics underground[78]—in this way they might penetrate 25 feet deep in a century, according to one estimate.[79] Other processes like erosion and landslides turn over dirt as well. So soil doesn't sit idly atop solid bedrock—it's humming with biological and geological activity, giving microplastics plenty of avenues to work deeper into the earth.

In doing so, microplastics transform the structure of the land. Lab experiments show that when added to dirt, polyethylene particles interact loosely with the soil matrix, meaning they keep to themselves.[80] But polyester and polyacrylic fibers act as nuclei around which bits of dirt clump, altering the arrangement of the soil. (These were pristine plastics, but keep in mind that bacteria in the plastisphere form a kind of glue, which attracts other microbes and potentially bits of dirt.) Soils high in microplastics, then, might erode differently than less contaminated land. All the polymers in the experiment also decreased the density of the soil—again, plastics are less dense than natural materials. But the particles decreased microbial activity, and by futzing with the structure of the soil, microplastics could influence how much water the soil retains or gives up to evaporation.

Any one of these changes, or several in coordination, might impact the denizens of the dirt. Scientists have shown that the soil-dwelling nematode worm *Caenorhabditis elegans*—better known as *C. elegans*, from its fame as a model organism for all kinds of lab research—suffers intestinal damage after eating microplastic.[81] Others have found the same for terrestrial snails, which ate 35 percent less when exposed to polyethylene and, yet again, broke the particles down into smaller pieces to be excreted.[82] However, another study on polyester fiber exposure to critters that work the same soil-processing gig as earthworms—a kind of isopod related to roly-polies (or pill bugs, if you prefer), mites, springtails, and tiny worms known as enchytraeids—found minimal harm, at least for a short experiment of a month.[83] (Antarctic scientists have also turned up microplastics in springtails there—the critters had been feeding on algae growing on a washed-up chunk of polystyrene foam.[84]) But the worms' reproduction decreased by a third when exposed to longer fibers, which hits at one of the many concerns about microplastic pollution: we need multigenerational studies on all kinds of organisms to see how they fare in the long term, because a monthlong exposure just isn't enough. As Edward Kolodziej points out, you don't need to kill a species to harm a species—messing with its reproduction is an efficient way to do that. Microplastics might also disrupt the gut microbiome of these critters, upsetting both their physiology and the way they process soil to create a healthy ecosystem for other animals and microbes to enjoy.

Birds, too, spread the microplastic plague from one soil ecosystem to the next. In 2018, scientists teamed up with Inuit hunters in the far north of Canada to collect northern fulmars and thick-billed murres from their breeding colonies.[85] These birds feed on fish, squid, and other critters in the thoroughly contaminated Arctic Ocean, and thus assume ownership of the microplastics their prey had eaten. The scientists dissected the fulmars and murres and examined the material in the last four inches of their intestinal tracts—essentially the birds' pre-droppings, what they would have excreted had they not donated their bodies to science—and calculated that in these two colonies, the animals discharge 50 million particles into the environment a year.

A separate survey of little auks—which are tuxedoed, like tiny penguins—looked at the other end of the digestive system in the gular pouch, where adults store food to regurgitate for their chicks.[86] (Scientists have tested the feces of actual penguins that dive for fish around Antarctica and found microfibers there too.[87]) Little auks dine on zooplankton in the Arctic, diving up to 160 feet deep to pick off their prey. That prey now includes microplastics: on average, the birds had 10 particles each in their gular pouches, which means adults are feeding petroplankton to their chicks. The majority of the particles were light colored, even though microplastics sampled from surrounding waters were mostly dark. So the auks seem to be mistaking microplastics for prey, as they prefer hunting light-colored zooplankton, which stand out against the darkness of the sea.

When these and other bird species fly around, they transport components of one environment into another. That's how plants spread: a bird eats a fruit and poops out the seed (conveniently packaged in nutrients) elsewhere. Northern fulmars, thick-billed murres, and little auks migrate up and down the coasts of North America, dining on seafood and excreting the digested material—and whatever microplastics had been in those animals—miles away.[88]

And let's not forget the indispensable pollinating insects. Entomologists have found microplastics in bee guts and have shown that feeding particles to the insects damages their digestive systems and makes them more susceptible to viral infection.[89] Plants, including of course our crops, rely on a diverse army of pollinators beyond bees, from beetles to ants, and even birds and bats.[90] As with plankton in the sea, microplastics might pile on other stressors—like neonicotinoid pesticides that are devastating populations of bees—to create a multifaceted problem for these pollinators.[91] Petroparticles might also act as vehicles for pesticides and pathogens, to say nothing of the many chemicals that come packaged in the plastic itself. Plus, pollinating mammals, birds, and insects have distinct physiologies, so a particular chemical might affect one group but not another—even related salmon species have wildly different reactions to tire particles. That and microplastic exposure changes at various stages of an insect's life. While butterflies and moths most famously transform from caterpillars into winged adults, bees also start life as larvae. In social bee species, workers feed a mixture of nectar and pollen to their young, dosing them with particles collected from the environment. Moths and butterflies gnaw on plants as caterpillars, and in the process hoover up microplastics that had settled on leaves. Then as adults, when they're landing on flowers and sipping nectar, they get particles that way too.

But if an insect ingests microplastics as a larva, might the particles remain as it transforms into an adult? One experiment found that for mosquitos, at least, it's possible.[92] Biologists fed fluorescent polystyrene beads to larval mosquitos, which develop in water, and watched as the particles persisted in the grubs as they grew. When the baby mosquitos developed into pupae—equivalent to a caterpillar shutting itself inside a chrysalis—still the particles endured. And as the pupal bodies completely rearranged themselves, metamorphosing into adult mosquitos, the microplastics went along for the ride. Thus a newly minted flying mosquito begins its life already tainted with microplastic. Synthetic particles, then, might transfer from water to land by stowing away in mosquitos, which themselves are eaten by all manner of bats and insects, further transferring plastics up the food chain. Obviously mosquitos aren't bees or moths or butterflies, but this persistence of microplastics between life stages may portend such contamination among the pollinators we rely on to propagate our crops.

## Food Fight

What microplastics are doing to crops themselves, that's a more complicated question.

Dannielle Green, the ecologist studying cigarette butts, and her team dosed soil with different kinds of microplastic, then planted perennial ryegrass—a popular kind of lawn in America—to see how it fared.[93] (They weren't particularly concerned about the vegetal health of your yard—ryegrass is commonly used in plant research.) They found that when exposed to microfibers or the biodegradable plastic PLA, fewer seeds germinated, and the grass grown with PLA also developed 19 percent shorter shoots. The addition of HDPE boosted root biomass, which sounds positive but may be the opposite: in response to stressors, plants expand their root systems in search of more water and nutrients. (In a separate experiment, Green found that the germination success and initial growth of ryegrass significantly declined when exposed to cigarette butts. It didn't matter if the butts were smoked or not, suggesting the plastic itself is noxious. The problem may be diethyl phthalate, a plasticizer known to be toxic to plants.[94]) In addition, HDPE made the soil's pH plunge, which could shake up the rhizosphere, or the interface between roots and the soil's microbial community.

This web of competing and cooperating plants and microorganisms, like bacteria and fungi, is extraordinarily complex and still largely mysterious.[95] Roots release goodies for microbes to feed on—proteins, sugars, amino acids—and in turn the microbes assist the plants, establishing a symbiosis. Most famously, mycorrhizal fungi form huge networks among the trees of a forest—some 90 percent of plant species depend on this partnership.[96] If one plant is attacked by a pest, it'll release chemicals to protect itself, but so too will another individual if the two are connected by fungi, even if the second plant remains untouched.[97] So the mycorrhizal network transmits alarms through the vegetation, as well it should: if the plants don't survive, neither will it.

Scientists have started to investigate microplastics' influence on mycorrhizal communities, at least in the lab, and the early results are mixed.[98] One team grew spring onions in soil laced with different polymers and found that, in general, the plastics decreased soil density while increasing the amount of water evaporated to the atmosphere. Polyester fibers doubled the biomass of the onion bulb and increased the amount of root colonization of mycorrhizal fungi by a factor of eight. PET, however, cut mycorrhizal colonization in half. Which lands us back at the perennial challenge of microplastics: this isn't a monolithic pollutant, but a family of polymers made of thousands of ingredients, any one of which might affect a given plant species—a cellulose acetate microfiber from a cigarette butt and a polyester fiber from a sweater are only similar in that they're fibrous. And sure enough, another experiment that exposed carrots to plastic fragments, foams, fibers, and films made of eight different polymers found a wide array of effects.[99] Films, for instance, increased the biomass of roots and shoots by 60 percent, whereas fibers only increased it by 6 percent. Soil became looser overall, which might explain the boost in root growth: the plant is more easily expanding through less-dense material. That might also explain why microbial activity dropped, as the transforming structure of the soil changes how oxygen diffuses and how well water trickles through and then evaporates out, but these effects were highly variable based on the shape and polymer of a particular particle.

Now, ryegrass and carrots and spring onions aren't what you'd call essential crops, but wheat certainly is. In one study, researchers grew wheat in soil spiked with bits of a biodegradable plastic mulch, with and without earthworms added to the mix.[100] The plastic—which is designed to be tilled into the soil after use—dramatically inhibited the growth of the wheat, perhaps due to some shakeup of the microbial community in the rhizosphere, as other scientists have found in separate experiments with biodegradable mulch.[101] The earthworms actually attenuated the effect of the plastic—when the animals weren't present, the polymer cut the biomass of the wheat grains we humans rely on.

But might crops absorb plastics in the soil, like they do water and nutrients, and incorporate them in these edible bits? To test this, Washington State University soil scientist Markus Flury and his colleagues at the Pacific Northwest National Laboratory recruited wheat and a common plant used in crop science, thale cress, as botanical guinea pigs. These species represent the two types of root strategies in the plant kingdom: wheat uses a fibrous root system, in which lots of filaments spread through the soil like a net, while thale cress uses a taproot system, in which smaller offshoots branch off from a thick main root—think of a carrot growing underground. Flury and his team grew seeds from each species in agar—a sort of scientific Jell-O—laced with fluorescent microplastic and nanoplastic beads.[102]

When Flury looked at the plants under a confocal microscope, which shines lasers that make the fluorescent dye in the plastics glow, he found particles had attached to the roots but hadn't penetrated them. So this is worrisome in that plastics might be accumulating around the roots we eat—carrots, sweet potatoes, radishes—but it's good news in that neither a fibrous nor taproot system seemed to uptake plastic into the plant itself, unlike how crops readily soak up nutrients like nitrogen and iron. "The plant has probably an incentive to take up an iron particle, whereas a plastic particle will not be used by the plant," says Flury.

This contradicts previous lab studies on wheat and other crops, like beans and onions and lettuce, showing that roots do take up plastics.[103] Over at ETH Zürich, analytical chemist Denise Mitrano took a different tack, tagging nanoplastics not with fluorescence but with the rare metal palladium. And instead of growing wheat in agar, she grew it hydroponically, exposing the growing plants to the "doped" particles. She could then track the nanoplastics as the wheat plants took them up into their roots and shoots. "We didn't let the wheat go to grains, so we don't know if the nanoplastic would eventually get into the food source, but it did go up further into the plant," says Mitrano. However, she didn't see any big changes in the physiology of the plants, like growth rate or chlorophyl production. "But we did see that it changed the root structure a bit and the cellular structure in the root, which would indicate that the plant was still under stress."

Basically, it's complicated, like testing how clothes shed microfibers in the wash. Scientists are exposing plants to different polymers in a range of sizes and concentrations, and they might grow their plants in agar, like Flury, because it's easier to count the particles accurately than in the complexity of soil, or instead opt for hydroponic systems, like Mitrano. So yes, in these hydroponic studies, particles infiltrated the plant. "The question then is, Does this really happen in the soil as well?" Flury asks. "And that's much more complicated. So you can demonstrate the potential of uptake with hydroponic systems, but in a real soil, where you have interactions of the plastic with the soil particles, where the water flow is not as fast as in a hydroponic system, the likelihood that you have uptake is much, much smaller." But by starting with these simplified agar and hydroponic systems, scientists can work their way up to testing microplastic and nanoplastic exposure in actual soils.

And while wheat and thale cress represent the two main root strategies, they don't speak for all the 400,000 species of plants out there.[104] "I think the more relevant question is, What happens with smaller particles?" Flury asks. "Forty nanometers was the smallest particles we used. And so if you go smaller, maybe there is more uptake potentially. But that we don't know." There's also the matter of concentration. Flury was using high levels of particles in this experiment, but if you go even higher, does that begin to overwhelm the plant? Even if the plastics don't infiltrate the tissues, might they accumulate in such numbers on the surface of the roots that they form a kind of blockade, interfering with the plant's ability to take up water and nutrients? In a separate experiment that exposed another kind of cress to high levels of plastic, researchers found that microplastics accumulated around the seed capsule, blocking its pores—which could interfere with water uptake—and significantly delayed germination.[105] Later on, the particles accumulated around the root hairs, as Flury had seen with thale cress and wheat. Later still, the researchers noticed the particles on the surfaces of the plants' leaves.

All these factors—the impacts on the physiology of plants and soil-dwelling animals, the structural changes to the dirt, the reshuffling of microbial communities, the persistence of tough-by-design polymers—make microplastic an unprecedented threat to soil ecosystems. And I'll emphasize that researchers are typically using virgin microplastics in their lab experiments, whereas particles out in the wild carry their own unique plastispheres—often including pathogens—which interact with native soil microbes in unique ways. As one team of scientists put it in a 2019 paper: "Microplastics are in some ways more reminiscent of invasive species than inanimate chemical toxins."[106]

Beyond crops, the animals on a farm are exposed to the microplastics from sludge, prills, and plasticulture. Working in Mexico's Yucatán Peninsula, scientists examined samples of soil, earthworm casts, and chicken droppings and then dissected chickens and removed the crop, where food is stored after it's swallowed, and the gizzard, a kind of stomach.[107] The researchers found minimal microplastic counts in the soil, but still more in the earthworm castings, and still more in chicken feces. This demonstrates a transfer of particles up the food chain: earthworms eat soil and accumulate microplastics and then get eaten by chickens, which accumulate the worms' accumulated particles. While the scientists didn't find particles in the chickens' crops, they did find them in the gizzards, a popular ingredient among locals for chicken soup. Given consumption rates of chicken, they calculated that the average Mexican could be ingesting 840 particles each year from gizzards.

Chickens may also get dosed with microplastics in their feed. To provide these animals with protein, feeds incorporate the same fish byproducts—the guts where petroparticles are accumulating—that farmed fish are eating in their feed.[108] So here we may have a particularly bizarre microplastic loop: we produce plastics on land, which end up in the sea as microplastics, which make their way up the food chain to larger and larger fish, whose guts end up back on land in livestock feed, potentially working their way out of a chicken's digestive system and into the muscles we eat. But this all demands more research, as no one has yet confirmed that fish guts are a source of microplastic in livestock feed.

Scientists have confirmed, though, that dairy and honey are contaminated with microplastics. A survey in Ecuador turned up an average of 40 particles per liter of milk and 54 per liter of honey.[109] The microplastics could have either been gathered by the bees in their foraging, or they snuck in during the packaging process. (Previously, scientists collected honeybees around Denmark and found microplastics stuck to their fuzzy bodies.[110] "When flying," the researchers write, "their bodies become positively charged with static electricity, so that when the bee lands on a flower, the pollen particles stick to their static-charged hair and the same happens with other microparticles in their environment.") Interestingly, samples of small-batch, hand-packaged honey in glass bottles yielded an even higher count of 67 particles per liter.

Spices, too, now come with a dash of microplastic. In Minneapolis, researchers gathered 12 brands of sea salt from around the world: Mediterranean, Himalayan, Hawaiian, among others.[111] They also made sure to get samples in different packaging, like cardboard, glass, and plastic, but every one of them was contaminated with microplastic, up to 23 particles per ounce of salt. This echoes previous findings from scientists in China, who searched 15 brands of sea, lake, and rock salt and found up to 19 particles per ounce.[112] According to an estimate by Evangelos Danopoulos of the University of Hull, you may be eating up to 6,000 microplastics each year in salt, but it really depends on the brand and where it came from.[113] "That's very interesting, because then you can connect it to the broader environmental contamination with microplastics," says Danopoulos. In contrast to sea salt forming in polluted ocean waters, if you mine salt from underground, you're gathering a mineral that formed long before humanity invented plastic. "So if we find some contamination there, then it means that it's not coming from the salt—it's coming from their manufacturing process or from packaging." Look closer at those particles, though, and you'll find clues written all over them: if a fragment or fiber snuck in during processing, it'll look less weathered. Really, you'd prefer to eat newer particles—an aged particle could have tumbled through the environment for years collecting microbes and pollutants.

The researchers in Minneapolis went looking through 12 brands of beer, too. All had been made with water from the Great Lakes and were packaged in either glass or aluminum containers. Yet all, like the salt brands, contained microplastic. Per liter of beer (about two pints), they averaged 4 particles, the highest tally being 16 from a brand made of Lake Michigan water. This contamination is not, mind you, an exclusive feature of New World suds: the beer purists of Germany are also swigging microplastic, according to a study that found up to 109 fragments per liter.[114] Which makes sense, given how plasticized water has become all over the world.

## Treacherous Waters

In 2020, Danopoulos cobbled together 12 studies that investigated microplastic in drinking water, representing Asia, Europe, and North America.[115] Six looked at a total of 40,000 liters of tap water and six looked at 435 bottles of water. The maximum microplastic count from the tap was 628 per liter—given a typical human's water requirements, that'd mean you'd be consuming up to 458,000 particles a year if you drink tap water alone. But that figure's puny compared to your exposure if you exclusively drink bottled water, where the maximum count was nearly 5,000 microplastics per liter, which comes out to more than 3.5 million microplastics a year. The particles come from the bottle itself and the shearing of plastic when you liberate the cap—they're different polymers, and scientists find both in bottled water—or they were in the water before it was ever poured in a bottle.[116] In any case, the World Health Organization (WHO) acknowledged in 2019 that drinking water of all kinds is laced with microplastic but stopped short of labeling it a threat to human health.[117] Instead, it called for more research.

Following a 2017 report that tested tap water from five continents and found North America's water had the most microplastic,[118] California passed legislation ordering its water board to begin investigating the pollutant.[119] More specifically, the board has to develop a methodology for testing water for microplastic, then run those tests for four years and report the findings to the public. Plus it'll answer the WHO's call and investigate how microplastic in drinking water might impact human health. "We're using the state as really the world's first case study in monitoring for microplastics," says Scott Coffin, a research scientist at the California Water Boards. "We're treating this as sort of an experiment, because at this point, we really don't know how concerned we should be about it being in our drinking water."

You might guess that I'm slightly more concerned about microplastics than most, and you'd be right. I live in San Francisco, which draws from Yosemite's Hetch Hetchy Reservoir, a source of water so pure, the state doesn't require it to be filtered before it comes out of my faucet. "No one has tested San Francisco tap water for microplastics," Coffin says. "But we believe that of all the places in California that we're going to find it, we're going to find it in highest quantities in San Francisco's, because there's no filtration at all." (It is, however, disinfected with chlorine and UV light—we're not barbarians.) Not that what's flowing to the rest of the state is pristine: the canal system that ferries water from the wetter northern parts to Southern California runs along a major highway and hundreds of miles of farmland, and we know how polluted those are with microplastics. "A lot of it's transported from stormwater runoff, but also just by dust," Coffin says. "So we're expecting to find a lot in the state water project." And anywhere else in the world, really, where drinking water is exposed to open air.

So Coffin is urging preemptive action, given that microplastics act as vehicles for heavy metals and pathogens, and the particles are themselves made of chemicals hostile to human health. (In 2022, California became the first state to adopt a microplastics mitigation plan, which will rein in single-use packaging and deploy better filtration systems for stormwater and wastewater.[120]) It's safe to say that Charles Goodyear and John Boyd Dunlop never worried about people consuming bits of their tires, which are made of ingredients never meant for human exposure—yet here we are. "Over about 80 percent are unknown to anyone but the manufacturer, including their toxicity," says Coffin. "So we're dealing with an iceberg and just being able to characterize the very tip of it."

Still, if you're drinking treated water, count your blessings. A third of humanity lacks access to a safe supply of drinking water, exposing these people to all kinds of diseases and potentially more microplastic.[121] If you're drawing your water from a plastic-polluted lake or river, you may be drinking a sort of plasticine soup—as bottles and bags float around baking in the sun, they're jettisoning little bits of themselves. Add to this the runoff from cities, bits of car tires and cigarette butts and fibers from our clothes, and a water source grows lousy with petroparticles. Indeed, a survey of 67 European lakes found the concentrations of microfibers to be four times greater in waters with high human activity than in more remote systems.[122] The densely populated Great Lakes are packed with particles: one study found over 12,000 microplastics and other anthropogenic particles in just 212 fish pulled from Lake Ontario—one unlucky animal had 915 pieces in its gut.[123] China's Three Gorges Reservoir, the largest in the country, tallies up to 50 microplastics per gallon of water.[124] Groundwater, too, is corrupted, as scientists are finding microplastics in aquifers, the particles likely permeating through the soil above.[125]

It's not just the water we should be worried about, but what we do with it. If a car tire is the ideal way to shed microplastics onto land, and plastic mulch is the ideal way to dose crops, preparing baby formula is the ideal way to load a liquid with particles, and for the same reasons: heat and friction.[126] Following the WHO's protocols for the safe mixing of formula, Trinity College Dublin materials engineer John Boland and his colleagues boiled several brands of bottles to sterilize them, filled them with 158-degree-Fahrenheit water, then added the powder.[127] They gave each bottle a shake, let the formula cool, and tested it for plastic.

Depending on the brand, the bottles shed between 1.3 million and 16.2 million particles per liter of fluid. Looking at the bottles under a microscope, Boland could actually see how the plastic had begun to flake away, layers of material eroding like the walls of the Grand Canyon. (A separate study of silicone teats found that disinfecting them with steam breaks down the polymer,[128] exposing an infant to an additional 660,000 microplastics by the time they're a year old. Gnawing on the teat liberates even more particles.) Formula consumption rates are higher in developed countries than in developing ones, so in North America, babies are drinking 2.3 million microplastics a day on average, and in Europe, 2.6 million. That'd mean a baby consumes nearly a billion microplastic particles each year. And that figure is based on the stuff that Boland managed to snag with his filters. "But actually, if we look at what goes through our filters, we see lots and lots of really tiny stuff," he says. "And that's what's really scary, because it's many *trillions* of nanoplastics. That's a size that in principle could get across biological membranes." No wonder, then, that a separate study found 10 times as many PET particles in infant feces than samples from adults, even though infants are a fraction of our size.[129]

Basically, when you and your children drink hot liquids, do consider how they're prepared. Takeout coffee cups, which are lined with polyethylene to keep the paper from disintegrating, release tens of thousands of microplastics and millions of nanoplastics—along with nasties like lead, arsenic, and chromium—into hot water in just 15 minutes, one experiment found.[130] Teabags are also made of plastic, and they shed particles when steeping.[131] Electric kettles inject tens of millions microplastics into a liter of water, but in another experiment, Boland discovered that if you boil enough tap water in a kettle, trace amounts of copper from pipes form a layer of copper oxide on the plastic—that's the dark gunk you see building up.[132] "That ultimately prevents the release of microplastics," says Boland. "That would happen also with baby bottles, actually, but parents are taught to scrupulously clean them." (Not that you shouldn't clean them. Just don't mix the hot formula in a plastic bottle.)

So hot water is a huge source of microplastics for infants and adults, but calculating how many particles we're consuming in total is fraught with uncertainties, given our diverse lifestyles. In 2020, though, a team of scientists gave it a shot, scouring previous studies that tallied microplastics in foods and beverages—salt, honey, beer, seafood, water—and married this with data on Americans' diets.[133] This didn't include fresh produce, meat, and grains, as the data on microplastic contamination there is still scant, nor did it consider the many packaged foods that Americans consume, which may be thoroughly tainted given the numbers we see in bottled versus tap water. The analysis accounted for just 15 percent of a person's caloric intake, and the researchers note that their tally may be a "substantial underestimate," as they were using the conservative estimates of particle counts in the various foods, and incorporated *recommended* consumption of the items, not what Americans actually put in their bodies. And, as always, these studies are missing the smallest of particles, which are even more numerous yet go undetected. Still, with all these caveats in mind, the scientists calculated that by eating and drinking these foods and liquids alone, the average adult American takes in around 50,000 particles a year, while children consume a bit less. (This paper published before Boland's revelation about baby formula.)

We know this is a substantial underestimate because we know what's coming out the other end of people. At the 2018 United European Gastroenterology meeting in Vienna, doctors for the first time reported microplastics in human stool.[134] The samples—taken from eight people in eight countries across Europe and Asia—suggest a person might excrete 1.5 million particles a year.[135] A second, larger survey in Beijing collected stool from two dozen people and found a similar amount.[136] In both studies, polypropylene and PET—commonly used to make utensils and food packaging—were the most abundant types of fecal plastic. So not only is wastewater flowing to the sea loaded with microplastics from our clothes and wet wipes but from our feces as well. Sea creatures like copepods and fish and whales then ingest those particles and release them in their own feces—a perpetual cycling of microplastics through digestive systems.

But that's not the whole picture of our microplastic exposure. In their limited inventory of microplastics in the American diet, those researchers also calculated another route through which we consume the particles: inhalation. The way they reckon it, we breathe in as many particles as we eat. Another team looking at microplastics in wild mussels found that you'd take in about 4,600 particles a year if you religiously ate the things, but urged us to focus not on what's in the mussels, but what *lands on* them: each year you might eat 70,000 particles that settle out of the air and onto your meals.[137]

Microplastics have sullied the air we breathe.

CHAPTER 4

# Breathe Deep the Plastic Air

Peering into a microscope in Janice Brahney's lab at Utah State University, I turn knobs to scoot around a white circle. If I pull back from the eyepiece and get my face close to the sample, I can make out a certain multicolored roughness, like sandpaper, the microscope's light casting shadows off the bigger particles. But with my eyes back in the device, that roughness turns into a galaxy of microplastics. There are chunks and microfibers long and short—reds and blues and pinks—each rendered in detail so fine, I can see a plastisphere of fungi speckled across one white shard. I tweak a knob to find black gobs of car tire and what Brahney reckons is a piece of a clothing label, since it's flat and two-toned, red on one side and white on the other. Nearby lies a microbead, perhaps washed from someone's face. And over here, a crystal-clear shard like an arrowhead.

Fragments and fibers and rubber all, they'd tumbled out of the sky and into Brahney's remote plastic catchers, like the one we'd hoofed up Beaver Mountain to see. By counting the particles in these samples, she can scale up her figures to estimate how much microplastic rains out of the atmosphere: in just 11 protected areas in the western United

States—a mere 6 percent of the country's total area—up to 8 million pounds fall each and every year, equivalent to 300 million bottles. Scaling that up even further to go nationwide (though keep in mind plastic will rain at different rates in different locations), that'd mean 5 billion bottles descend on the US as microplastic annually.

The usual caveat applies here: Brahney was sampling particles down to four micrometers, so this is an undercount of what's deposited. The atmosphere is infested with even more nanoplastics, since microplastics continue to break apart in the air, and the smaller a particle is, the more easily it takes flight and the farther it flies. Scientists working at a research station 10,000 feet up in the Austrian Alps—about the same altitude as Beaver Mountain—determined that a minimum of 19 billion nanoplastics are deposited there per square foot of snow every week.[1] That'd mean if you stood atop this remote mountain for an hour, 100 million nanoplastics would settle on your head and shoulders. The researchers further modeled the atmosphere at the time of sampling, showing that major sources of particles were European metropolises like London, Paris, and Amsterdam—the bigger the population, the more nanoplastics thrown into the sky. But they also reckoned that 10 percent of the nanoplastics had blown from more than 1,200 miles away. Another analysis of an ice core from Greenland found substantial nanoplastic deposition going back to 1966—that was the deepest the core went, so it's not like the dusting started that year—the most common polymer by far being the most commonly *used* polymer, polyethylene, but also bits of tire.[2] (The scientists also found nanoplastics in ice samples from Antarctica, but most of those were likely scavenged from surface waters, as Melanie Bergmann found with Arctic ice.)

Earth's atmosphere hasn't just grown saturated with plastic—plastic is now a fundamental component of the air. Brahney's been finding that concentrations don't decrease the higher up a mountain a sampler is, which you might expect as you're moving away from human activity, but actually *increase*. "The higher elevation you get, the more plastic deposition we see," says Brahney. "Which is an indication of long-range transport."

We'd expect nothing less of the atmosphere, which transports bits of dirt vast distances. Dust from the Sahara regularly blows into Europe, turning skies and snows orange. It'll even fly clear across the Atlantic, and in fact the forests of South America rely on the grime for phosphorus: each year 61 billion pounds of dust reaches the Amazon basin, enough to fill 100,000 semitrucks, and another 100 billion pounds reaches the Caribbean.[3] Wildfires, too, seed the atmosphere with particulate matter—smoke from the West Coast of the US reaches the East Coast, and even Europe if the conditions are right. And volcanoes launch plumes of particulates high enough into the atmosphere to encircle the globe. Plastics are half as dense as these natural materials, so they're easily kicked into the atmosphere and floated around the world. Microfibers in particular are well-shaped to take flight since they're long with a lot of surface area—they're basically airplane wings.

And just like airplanes, the things can get anywhere. A special atmospheric sampling aircraft once gathered microplastics 11,500 feet high.[4] The particles are in snow cores from Iceland's Vatnajökull ice cap, 40 miles from the nearest urban area (Höfn: population 1,800), and on remote Mount Derak in Iran.[5] At the tippy top of the French Pyrenees, researchers tallied 34 microplastics landing on a square foot of ground each day, having likely blown from Barcelona 100 miles to the south,[6] while on Baffin Island in the Canadian arctic, others found more than six times that, the vast majority being fibers.[7] A rain gauge in Jakarta accumulated four times the microplastics in the rainy season than the dry—falling raindrops scavenge particles from the air, like freezing ice pulls them out of the Arctic Ocean.[8]

No matter how remote a location or how high the elevation, Brahney and other scientists are finding startling amounts of microplastic there.

So, I ask Brahney, is there any reason to believe there's a place on Earth that hasn't been corrupted?

"No," she says quietly, shaking her head. "No."

## Wind Advisory

Microplastic pollution is so out of control, and has been out of control for so long, that particles of diverse origins have mixed in the atmosphere—what I saw under the microscope was a hodgepodge of fibers and fragments and microbeads and tire particles. Put another way: the microplastic community in the air has been homogenized, more so than what you'd find in sludge, largely microfibers from laundry and particles from cosmetics. So Brahney can't look at a sample and say, *OK, this bit came from just down the road, and this bit came from the nearest city*, because they're coming from everywhere, mixing into a synthetic dust in the sky. New particles mingle with old, PET with polystyrene, wisps of film with wing-shaped fibers.

What Brahney can do is work with atmospheric scientists to model the likely source of what's falling into her samplers in the remote US. If they have microplastics collected on a given day, and they know what the atmospheric conditions were at that time, they can trace the journey the particles took to get there. So instead of assuming a tire particle came from the nearest highway, the scientists more accurately track it based on variables like wind patterns. Because all this pollution has been mixing together since plastic production took off in the 1940s, it isn't so easy as just saying a microfiber must have come from the sludge of a nearby field, because it may have originated thousands of miles away.

In 2021, Brahney and her colleagues published the bleak results of their atmospheric modeling: at any given time, over 2 million pounds of microplastics are blowing above the western US.[9] Of that, 84 percent fly off roads—tire and brake particles but also ground-up roadside litter like plastic bags and bottles—not so much from big cities, but from just *outside* of them. Urban centers contributed just 0.4 percent of the microplastics in the atmosphere, because tall buildings block winds from scouring streets, plus cars can't travel as fast as they can on the open road. "Cars driving on roads provide that mechanical energy to move particles really high into the atmosphere," says Brahney. All that dust that you see kicking off cars on the highway? A good amount of it is microplastic, and a good amount of it is getting flung high enough to become atmospheric. Gravity eventually pulls some of the particles back to Earth, while others get caught up in raindrops and trickle into rivers and lakes.[10]

These road microplastics turn out to be surprisingly complex. Another group of scientists discovered that "tire-core particles" act as nuclei, gathering a crust of particulates jettisoned from brakes and the road itself.[11] That's because rubber is flexible and rough, making it sticky, in a sense. "In addition," the researchers write, "the particles display a rounded cross section, which allows them to roll over the road surface easily, thereby collecting other road-dust particles like a rolling snowball." Just as a microplastic adventuring around the ocean gathers a plastisphere, so too do these tire-core particles gather a complicated community of hangers-on. The researchers identified hitchhiking minerals from the road itself like quartz and gypsum, toxic metals like barium from brake pads, and road salt, plus other elements like tungsten, sulfur, and chlorine. This in addition to the many nasties from the tire material itself, like lead and cadmium.

These researchers further found that slower roads generate tire-core particles with more entrustment, which might seem counterintuitive—you'd assume that faster traffic would lead to more wear and tear on the road and cars, thus producing more particulate matter to glom onto the tire nuclei. But they reckoned that faster traffic could provide enough energy to fling a tire particle into the atmosphere before it has the chance to gather a thick crust. On slower roads, cars only provide enough energy to make the particles lazily tumble down the street, thus initiating the snowball effect. So in dense urban areas, where traffic is much slower than on highways, "the tire-core particles are exposed to repetitive cycles of slow roll-over processes with long-term contact between the rubber and road dust, which results in a high degree of encrustment," the researchers write. "On roads with high velocities (e.g., motorways), however, a more efficient removal of material from the road surface due to vehicle-induced turbulence is observed."

But here, a potential solution: engineers can exploit these physics to intercept tire microplastics before they reach the road. A startup called the Tyre Collective has developed a sort of microplastic magnet that attaches to the underbody of a car and hovers around the back of a tire, like a bracket. As the tire rolls along, it generates both friction and microplastics. "So as they're falling off, they have a slight charge on them," says Siobhan Anderson, cofounder and chief scientific officer of the Tyre Collective. "And we are able to adapt an electrostatic precipitator, which is essentially something that generates an electric field, and using that we can direct the particles onto a collection plate." Interestingly, Anderson has seen that tire nucleation effect on the plate. "If you already have some rubber on the surface, it will help other rubber particles as they're flying past adhere to that." Anderson says that in lab experiments, the Tyre Collective's device collects 60 percent of emitted tire microplastics, which accumulate in a container. So in the future, a mechanic might change your oil, swap out air filters, and empty your microplastic catchers—the collective is already talking to carmakers.

The urgency is real, as roads spew microplastics unchecked. One atmospheric model has shown that smaller tire and brake particles readily take to the sky and stay aloft for a month, more than enough time to blow across continents and oceans.[12] Indeed, the modeling produced a world map in which few places *don't* get fallout from faraway roads.

The Arctic is heavily dusted with particles from Europe, both from cars and microfibers from clothing, but don't just take the modeling's word for it: an expedition to the region found 14,000 particles in a liter of snow—the grime from distant cities.[13] Over in the tropical forests of southwestern China, scientists found hundreds of microplastics per pound of soil, the particles having drifted down through the canopy.[14] Still others working in microplastic-dusted lakes on the remote Tibetan Plateau found up to 1,200 particles per pound of sediment.[15]

This is not all to say that cities themselves aren't teeming with microplastics—it's just that urban roads tend to better hold on to the petroparticles they generate, plus at the same time metropolises are getting microplastics from the sky. In Paris, collectors trapped up to 26 microplastics per square foot a day—perhaps 38,000 pounds of fibers alone deposited on the city every year—about the same as others have found in Hamburg and the Chinese city of Dongguan.[16] In London and Ho Chi Minh City, the counts were triple that.[17] Where plastic is incinerated, the fallout is even worse: in low-income countries—particularly island nations with little space for proper landfills—90 percent of waste is either thrown in open pits or burned, creating a rising column of hot air that flings both microplastics and toxic chemicals into the atmosphere.[18]

Every urban area, then, is dusted with microplastic generated both locally and flown in from afar.[19] In the Iranian city and county of Asaluyeh, researchers found loads of petroparticles in urban dust, but they also took air samples and found the microplastics there to be finer and more fibrous.[20] The likely source? Pedestrians walking down the street shedding microfibers, which settle in the dust before the wind or a car or a footstep kicks them up again. The scientists calculated that at these concentrations, an adult and child could inhale 50 and 100 microplastics, respectively, walking down the street each day, a child's face being closer to the ground, thus increasing exposure to dust. Over in

Surabaya, Indonesia, researchers found even more microplastics floating in the air and likewise saw that fibers dominated the samples—the more congested the street, the more suspended particles.[21]

But to return to Brahney's modeling. In addition to the 84 percent of microplastics coming from roads, she traced another 5 percent to soils—the microfibers from sludge and plasticulture, to be sure. But the remaining 11 percent come from the ocean: having long received microfibers from wastewater, the seas are burping them back onto land.

It was actually Deonie and Steve Allen, the scientists we met in the first chapter, who discovered this phenomenon.[22] The thinking used to go that the ocean served as a microplastic sink: rivers and wastewater would carry the particles to the sea, and that's where they'd stay. But the Allens revealed that no, the ostensibly fresh sea breeze you enjoy at the beach is laced with microplastics.[23]

When a bubble ascends from the ocean depths, it gathers debris along the way, like viruses and bacteria and bits of dead animals. Once it reaches the surface and pops, it flings that material into the air. As water rushes in to fill the void left by the bubble, a vertical jet shoots still more aerosolized material into the air. Between this action and crashing waves, the air above the water gets loaded with particulates, which is a perfectly natural process, as fog forms as moisture around such objects.[24] Such oceanic aerosols are why you can smell that characteristic ocean scent—the same thing happens with beer foam, as it happens, which propels aromas into your nose.

Well, at least this *was* a perfectly natural process, until humans started loading the seas with microplastics, which hitch a ride on bubbles, ascend to the surface, and fire into the atmosphere. The Allens calculated that in this way, 300 million pounds of microplastics blow out of the oceans and onto land each year. "And this isn't just the sea—it works in freshwater environments, anytime you've got a bit of a wave or you've got some turbulence," says Deonie Allen. "Raindrops do the same thing." That means any lake or river polluted with microplastics—that'd be all of them at this point—emits petroparticles to the air when it rains, each drop crashing into the water and ejecting tiny droplets that rise in a crown shape. The droplets, which are now about 15 percent raindrop and 85 percent surface water, can rise several feet in the air, where the storm's winds will pick up the spray and loft it into the atmosphere.[25] (This process also means marine animals without gills, like whales and dolphins and sea turtles, are inhaling microplastics when they come to the surface to breathe—whales in particular take in great lungfuls before diving thousands of feet deep.[26] As these animals splash around the surface, and as whales blow water vapor out their blowholes, they launch still more microplastics into the atmosphere.)

The more polluted the body of water, the more microplastics it spits into the air. The Allens have collected air samples at the Pic du Midi Observatory in France, almost 10,000 feet above sea level, and determined that microplastics there had flown from as far away as North America—2,800 miles on average, the longest distance being 6,300 miles.[27] But their modeling also exposed an outsized contributor of these particles: the Mediterranean Sea. All those microfibers washing into the Mediterranean and gathering in sediments don't stay in the sea for good—they're coming to the surface and taking to the air. In this part of the world, particularly along the sea's African coastline, blazing temperatures create rising columns of air that send microplastics above the clouds to catch atmospheric highways—the same ones airliners use—where they travel thousands of miles for weeks at a time before falling back to Earth once more. "We didn't go in there expecting to find that," says Deonie. "We were expecting it to move a long way, and we were expecting to be able to find plastic, but we didn't expect to be able to see it coming out the sea." The Mediterranean isn't so much a sink of microplastic, where particles get stuck forever, but a substantial source of microplastic into the atmosphere.

Back in the US, Brahney's modeling found that sea spray could be responsible for a tenth of the atmospheric microplastics in the American West. But microplastics are also going the other way: if scientists in the Arctic are finding particles that have blown from Europe, that means they're blowing from all the land masses into the ocean, only to be belched up and thrown onto land once more. "That's a substantial shift in thinking," says Steve Allen. "We assume that it's coming out of the wastewater treatment plants and washing through river systems, but actually there appears to be a hell of a lot that's just blowing off the land."

By Brahney's calculations, we may have reached a point where there's now more microplastic blowing out of the sea onto land than the other way around. "That is, the continents emit less plastic to the ocean through the atmosphere than the ocean gives us back through the atmosphere," she says. "There's just so much plastic out there. Anything that we produce in any given year is just a small fraction of what we've already put into the environment that ultimately ends up in the ocean."

And so a "microplastic cycle" comes into view. When you wash your clothes, synthetic fibers flush to a wastewater treatment facility. From here, they'll either be pumped out to a body of water, only to be burped back onto land, or sequestered in sludge and spread on fields, only to be scoured by winds and taken up into the atmosphere. They'll blow hundreds, perhaps thousands of miles and fall on a mountain or in a rainforest or Tibetan lake. Some will drop into the sea, where they'll cruise around on currents gathering plastispheres, maybe even ones thick enough to sink them to the seafloor. Or more exciting still, they'll get eaten and sink in a fecal pellet, then get eaten by something else—moving up and down the water column before they're captured by a bubble and flung into the atmosphere to blow onto the land whence they came.

It's a journey that petroparticles have been making unnoticed for three-quarters of a century. The Allens have sifted through "ombrotrophic" peat in the remote Pyrenees for atmospheric microplastics and found yet another echo of Jennifer Brandon's exponential sedimentary particle counts.[28] Peat is wet plant material that resists decay, so it accumulates layer after layer, and this particular variety is ombrotrophic because it gets water, nutrients, and pollutants not from rivers but exclusively from the atmosphere. That makes each layer of material an excellent indicator of what had fallen out of the sky at a particular time. Digging down to peat that'd been exposed to the air between the 1940s and '60s, the Allens found negligible amounts of microplastic. But as the decades went on and the atmosphere grew more burdened with petroparticles, so too did the peat—now it's 500 microplastics deposited per square foot each month. Overlaid with the rising number of particles in sediments, fishes, plankton, and greenhouse-adjacent seabeds, peat reveals a hitherto invisible tragedy: as humanity produces exponentially more plastic, every corner of the environment grows exponentially more plasticized.

What scientists don't yet have a handle on is how microplastics might influence the atmosphere itself. Like fog, a cloud forms when water vapor gloms onto nuclei like dust, sea salt, and bacteria. Snow works the same way when water freezes around nuclei. In the lab, at least, scientists have shown ice nucleation around microplastics and nanoplastics in special chambers that replicate the conditions of the atmosphere, and others have found the particles in snow.[29] But in the wild, these particles accumulate bustling plastispheres, which may change how they interact with water. "What are they actually doing in the atmosphere?" asks atmospheric chemist Laura Revell of the University of Canterbury in New Zealand. "If they are getting into clouds, rainfall patterns can change due to a cloud being perturbed by lots of aerosols."

Two things could be happening here. One, microplastics might indeed act as nuclei, thus altering the formation of clouds. The more nuclei, the more water droplets you get, but the smaller those individual droplets turn out, as there's only so much water to go around. Extra nuclei also brighten a cloud, changing its albedo and bouncing more of the sun's energy back into space, thus cooling the local area. But that could mess with the weather too. "Anytime you're changing temperature patterns in the atmosphere," says Revell, "that'll also lead to changes in circulation and precipitation."

And secondly, microplastics may influence the climate at large. Aerosols in general can substantially alter the way energy is both bounced off the planet and trapped against it. "They can scatter sunlight back to space," says Revell. The ash from the eruption of a large volcano, for instance, blocks the sun's energy, thus cooling the planet. "Or depending on the type of aerosol and where they are in the atmosphere, they might also absorb radiation that's emitted by the Earth."

Revell's early modeling hints that atmospheric microplastics may already be doing a little bit of both, the particles reflecting solar radiation but also trapping heat against the planet.[30] At the moment, the former effect is probably winning out, thus ever so slightly cooling the climate, but we've got some caveats here. Because it'd be impossible to account for the extreme complexity of airborne microplastics—different polymers and plastispheres and colors and shapes and sizes—Revell instead broke the particles into two main groups, fragments and microfibers, and calculated the general optical properties of each, like how well they scatter sunlight. This was for *nonpigmented* microplastics, but what's actually up there is a technicolor dust of petroparticles, and the darker a particle is, the more of the sun's energy it absorbs. Also, in the modeling, Revell used a single concentration of atmospheric microplastics, but scientists are finding that concentrations vary quite a bit depending on where you're looking—that and the population of particles in the atmosphere is growing day by day. So taking into account the colors of atmospheric microplastics and their growing numbers, they could well end up warming the climate. "Even if we were to just turn off plastic production today," says Revell, "we would still be dealing with this problem for such a long time to come."

## Troubles at Home

Take a look around. If you're on a bus or train, you're sitting on a plastic seat surrounded by people in synthetic clothing, all of it shedding particles as they move. If you're on the couch or in bed, you're sunk into the embrace of microfibers. The carpet underneath you is plastic, as is the coating of a hardwood floor. Curtains, blinds, TVs, coasters, picture frames, cables, cups—all of it's either wholly plastic or coated in plastic. Whereas the takeover of packaging was a conspicuous revolution for plastic bags and bottles, the material's infiltration of every other aspect of our lives has been a quiet coup. While scientists like Revell, Brahney, and the Allens have been untangling the complex dynamics of microplastics in the atmosphere, others have turned their attention to how the omnipresent plastic products around us are bastardizing our indoor air. In 2015, researchers sampled the living rooms of two apartments in Paris, each home to two adults and a child, as well as a university office where three people worked.[31] They only sampled air when people were present in the rooms, both at a height of about four feet to gather what the subjects are breathing, and a half inch off the ground to determine the deposition rate of dust. The researchers also took samples from vacuum cleaner bags the occupants had used in the two apartments.

In one apartment, they counted up to 14 fibers floating in a cubic yard of air, 12 in the other, and 45 in the office. Based on the number of particles they caught near the floor, the researchers calculated that up to a thousand fibers are deposited per square foot each day, which matched with the number of fibers they found in the vacuum bags. (A separate study in 2019 that collected dust from Chinese homes found almost the exact same amount, as it happens, while another in New Jersey found twice as many particles.[32] The nations of the world may have their differences, but microplastic pollution is not one of them.) Overall, two-thirds of the fibers they tallied were made of natural materials like cotton and wool, while the remaining third were plastic. Polypropylene fibers were particularly prominent, and indeed one of the occupants clued the researchers in on the fact that they'd been sampling a room adorned with a large polypropylene carpet.

Moving over to the West Coast, another team tested indoor and outdoor air on the California State University Channel Islands campus.[33] They found a similar concentration of microfibers suspended in the air indoors and discovered that microplastic fragments had become airborne as well. The more foot traffic the area had, the higher the microfiber count. "Fibers from the synthetic clothing of students and staff passing through could easily contribute to microfiber loading of the inside air," the researchers wrote in a paper. They collected more than six times the number of microfibers indoors as they did outdoors: with little airflow inside, the particles suspend in the air, waiting to be breathed in, whereas outdoors plentiful airflow dilutes the particles.

We are all, then, like Pig-Pen from the *Peanuts* comics, who swirls with a perpetual aura of dust, only we're depositing our microfibers wherever we go. As you abrade a synthetic fabric—putting it on or walking around in it or sitting on the couch—its fibers "fibrillate," meaning that instead of always breaking neatly in two, fibers also shed clones of themselves, known as fibrils. Under the microscope, the fiber looks like a giant mother surrounded by tiny, curled-up offspring. One experiment found that abrading an ounce of fleece produced 60,000 microfibers, but also 170,000 fibrils that were significantly shorter and thinner than their parents, and therefore more liable to get suspended in the air around us, à la Pig-Pen.[34]

To be clear, that was done with a standard testing machine the textile industry uses to abuse new materials, not on humans walking around a room. To test this directly, scientists recruited four volunteers to move around a space wearing four different kinds of synthetic garments. After counting the microfibers from petri dishes left in the room, they arrived at a stunning figure: each year, you might shed a billion polyester microfibers into the air just by moving around, which would explain why all these studies find so much microplastic deposited on floors.[35] This is based on those four specific garments, though, so your results may vary—if you wear a lot of cheap, fast fashion, you may be shedding more.

Another study in 2020 confirmed the findings from these indoor air surveys, and on longer timescales to boot.[36] In Shanghai, researchers sampled a dorm room, an office, and a corridor of a lecture building. In the dorm, they counted up to 7,000 particles deposited per square foot of floor each day, 1,200 in the office, and 1,600 in the corridor. Like the Paris study, they found that about a third of the particles were plastic, while the rest were natural fibers like cotton. But because these researchers were sampling continuously for three months, they could chart how deposition rates changed day to day: the particle counts tripled on weekends in the dorm and doubled on weekdays in the office, while the counts in the corridor remained stable over time. That tracks with the behavioral patterns of students spending more time at home over the weekend but more time in classrooms and offices during the week. The researchers also futzed with the air-conditioning in the dorm and found that having it on at any speed significantly increased the number of microfibers deposited, as vigorous air flow picked up particles that'd settled on furniture. AC units themselves both capture and release additional microplastics: they're snagging particles when air passes through their filters, sure, but those filters are also made of plastic that sheds fibers, which are then blasted around the room in the cold air.[37] The flow of human bodies through a room or hallway generates still more airflow, churning up microfibers that had settled on the floor and other surfaces. That's why the air in busy rooms consistently tests for more microplastics, says Paris-East Créteil University environmental scientist and chemist Rachid Dris, who did the Paris study. "We notice always that it's the ones where there are more people coming and going, we will have higher concentration than the ones where there isn't lots of movement. And that is probably due to this resuspension effect."

Scientists aren't just finding plenty of textile microfibers in indoor dust—polymers like polypropylene and polyester and polyamide, from clothes and rugs and couches—but polyvinyl microplastics (PVC is polyvinyl chloride) as well. Neda Sharifi Soltani, an environmental scientist at Macquarie University in Sydney, led a 2021 study of indoor air in Australian households, where she found similar microplastic deposition rates as Dris did in Paris.[38] But in households without carpet, she found that polyvinyl, a polymer used in linoleum and wood flooring finishes, was the dominant microfiber. Polyvinyl was twice as prevalent, in fact, as in homes with carpet. "When I look at the samples under the microscope, it is really, really surprising—lots of fibers we are exposed to every day," says Soltani. (Religiously vacuuming, then, will go a long way in reducing microplastics in your home, whether you have carpets or not. Just be careful when disposing of the dust so you're not flinging the particles into the air again. Sweeping will be less effective, since that mechanical action resuspends some of the microplastics.)

So then, how many of these particles are we breathing? We've got these consistent tallies of microplastics swirling in the air and collecting as dust on the floor. We know how much air a typical human breathes each year, and we know that people in high-income countries spend 90 percent of their time indoors, where microplastic pollution is far worse than outdoors.[39] We've also got a good amount of complicating factors, of course, like how many sources of microfibers are in a room and how big the room is and what the airflow's like. But we've got enough data to ballpark it: by Soltani's estimate, we inhale 13,000 microfibers a year. Other scientists' estimations have quadrupled that figure. Another rather quirky experiment used a mannequin with mechanical lungs to calculate that an adult might inhale 272 particles a day, or 100,000 in a year.[40]

But in 2021, Fay Couceiro (who proposed the traffic light system for textiles in the first chapter), claimed a much, much higher estimate. She visited a home and gathered airborne microplastics with a pump (sans mannequin) that approximated human inhalation, then used micro-Raman spectroscopy—a particularly sensitive version of the microplastic-counting technique—to detect particles between 1 micrometer and 10 micrometers, the size of a single bacterium. Her tally: we're inhaling up to 7,000 microplastics a day, or 2.5 million annually. The average human takes 20,000 breaths a day, which would mean that with every third breath you inhale a microplastic. Couceiro did this experiment in a bustling home with two children, so there was ample opportunity for particles—especially at the tiny size she was looking for—to resuspend in the air. "I have kids myself—I've seen what my kids do," says Couceiro. "They jump up and down on the bed and hit each other with pillows. You can *see* a lot of particles in the air when you walk into the room. And that was what I wanted to show, that if you were in that kind of environment, you would be breathing in a lot more than perhaps we thought we were."

Physiologically speaking, though, children likely inhale fewer microplastics than adults because they're smaller. But from a *behavioral* standpoint, they may be inhaling more: kids are up to the aforementioned shenanigans, and toddlers spend a lot of time crawling on the floor, where microfibers accumulate—thousands of particles per square foot each day. That finding I mentioned before, that infant feces tests for 10 times the amount of PET—a.k.a. polyester—as adults, probably has something to do with children chewing on clothing and spending so much time on the ground in contact with microfibers, in addition to their formula being laced with microplastics. Toddlers also gnaw on plastic toys and may ingest particles that way too.[41] Crawling children, plus adults and pets shuffling around, will stir up particles, resuspending them for everyone in the room to breathe.

And the usual caveat applies here: even with micro-Raman spectroscopy, researchers can only quantify particles down to a certain size, so the smallest are escaping detection. Actual plastic particle counts in indoor air and dust are likely much higher—consider the millions of nanoplastics falling on your head if you're standing *outside* in the Alps. Given how difficult and expensive it is to test for nanoplastics, though, that remains an assumption. "But it's a very reasonable one," says Dris. So unless you work in a factory making synthetic textiles, the most polluted place you frequent may be the room you're sitting in right now. (Wearing face masks during the COVID-19 pandemic has had contradictory effects here.[42] The synthetic material keeps out both the virus and microplastics swirling in indoor air but also sheds fibers for us to inhale.[43] Don't get me wrong—that is far and away a better outcome than eschewing masks and getting COVID. But disposable masks are now all over the environment: a study estimated that in 2020 alone, 1.5 billion masks may have entered the oceans, and another found that one of those masks releases 1.5 million microplastics as it decomposes.[44])

All of it's coming from the towels we wipe our hands with and the clothes we wear and the couches we plop down on and the carpets we tread—just look at the sunlight coming in through a window and you can watch airborne microfibers dance in the beam. (I also find them—as I write this sentence—stuck to the lenses of my glasses.) Any plastic product you interact with, be it a trash can or coffee maker or lamp, is jettisoning little bits of itself as it ages. Rub against lacquered furniture, and off come microplastics. Cutting open a single-use plastic bag produces particles as the material shears, and ripping one open adds extra energy to fling microplastics into the air.[45] Same goes for breaking the seal on a plastic bottle cap. Whenever you run your clothes dryer, microfibers tear loose and accumulate in the lint trap. When you clean out the filter to keep your house from burning down, you're holding concentrated microplastic, which flies into the air of your laundry room.[46]

And into the environment too. In one clever experiment, scientists dried pink polyester fleece blankets in two homes after a fresh snowfall, allowing them to easily scour the area around the dryer vent for fibers of the same color.[47] They took each sample from a square foot of snow at different distances out to 30 feet, both laterally and straight back from the houses, 14 plots total in each yard. They found an average of 400 fibers per plot in one yard and 1,200 in the other. The most fibers accumulated closest to the vents, but the researchers did find that many had made it out to 30 feet. And they weren't sampling for nanoplastics, which would more easily slip through the lint filter and take to the air. A separate study that tested polyester clothes estimated that your dryer could be emitting 120 million microplastics into the outdoor air every year.[48] And keep in mind that the air coming out the dryer vent is hot, so it rises and boosts the particles into the atmosphere.

So we have yet another source of microfibers into the environment: high heat and friction in a dryer conspiring to brutalize the plastics in our clothes. And as more people around the world ascend into the middle class, more washers and dryers are coming off production lines. Which is not to say line-drying is any better, or worse, than using a machine—the synthetic fabrics battered both by winds and UV radiation, as plastic mulch flaps in a field—it's just that no one has yet quantified the microfibers released. A clothesline has no filter, while a dryer's lint filter does a so-so job of snagging microfibers to keep them out of outdoor air, though obviously many are slipping through. And when you remove that accumulated lint and throw it in the trash, that's no guarantee the fibers won't take to the air at some point in the waste management process.

So, to add it all up: how many total microplastics might we be consuming a year by eating, drinking, and breathing? Every human will differ, and there's no way of knowing *exactly* how many particles go into your body. But according to a 2021 study, which aggregated all kinds of data about known human exposure to microplastics, the average child takes in 553 particles per day, or 202,000 per year.[49] For adults, it's 883 per day, or 322,000 per year. They were only able to account for a fifth of food consumption, given the lack of data on so many products. And once again we have a discrepancy here, because stool samples suggest we excrete 1.5 million particles a year, and if Fay Couceiro is correct, we may be inhaling millions more.

Whatever the number ends up being, it's going to be big, and day by day it gets bigger as the production of plastic accelerates. Scientists aren't waiting around for a definitive answer—they are now racing to understand what microplastics are doing to our bodies.

## Breath of Unfresh Air

Breathe deep. Air passes through your nose, down the back of your throat, and into your windpipe.[50] The air enters the bronchial tubes, one that leads to the left lung and another to the right. Once in the lungs, the air flows into smaller passages known as bronchi and then into even smaller bronchioles, which terminate in air sacs. These alveoli are packed with capillaries, in which blood absorbs oxygen from the air.

As long as humans have walked the Earth, we've breathed dirty air, all kinds of bacteria and fungi and viruses, plus smoke and dust particles, none of which are good for our lungs or bodies at large. To expel these intruders, the airways of our lungs are lined with a mucociliary escalator, a dense mat of whiplike projections called cilia that beat back and forth to propel a layer of mucus. This snags any foreign objects before they can reach the alveoli and shuttles them back where they came from. So when you inhale smoke and you cough, that mucus is the product of the mucociliary escalator. You swallow that gunk, thus sequestering the smoke particles—you'd rather have them in your digestive system than your respiratory system, where they're liable to cut off your oxygen supply if enough particulates accumulate in the alveoli.

But like a person can sprint down an escalator that's going up if they try hard enough, intruders can sometimes run against the mucociliary escalator and end up in the alveoli. That's because the escalator peters out the deeper you go in the lungs, since having gobs of mucus in the alveoli would interfere with their ability to absorb oxygen from air. Also, trauma like years of cigarette smoking can damage the mucus membranes in the lungs, crippling the escalator. But even in healthy lungs, certain particles get lodged in the airways, most notoriously asbestos, a mineral used to fire-proof building materials. These tiny, needlelike fibers penetrate deep into the lungs and stay in the lungs, mucociliary escalator be damned. Normally if a particle won't budge, the body's immune cells attack and dissolve it, which works fine for bacteria and viruses. But not asbestos fibers. They persist, leading to irritation in the lungs and all kinds of maladies, including cancer and asbestosis, a chronic disease. "We do biopsies of people who've worked in asbestos factories—you actually can see the asbestos fibers," says Albert Rizzo, chief medical officer of the American Lung Association.

Doctors are finding the same for long, skinny microfibers, which like asbestos fibers may be perfectly shaped to penetrate deep into the lung.[51] Back in 1998—six years before the term *microplastic* was coined—researchers at the Roswell Park Cancer Institute reported troubling patterns in the lung tumors of 114 patients.[52] They found either cellulosic (cotton, for example) or plastic microfibers, or both, in 99 of them. Of the 33 malignant lung specimens, 32 contained fibers. "Most fibers exhibited little or no deterioration," the scientists noted in a paper, "and this observation supported our premise that the inhaled cellulosic and plastic fibers were bioresistant and biopersistent." Like with asbestos, not only were the lung's clearance mechanisms failing to evict the fibers, but the foreign objects weren't breaking down. "Accordingly," they concluded, "it would be reasonable to postulate that inhaled biopersistent cellulosic and plastic fibers, particularly those that contain mordants, dyes, and various chemicals, may contribute to different pulmonary diseases, including lung cancer."

Scientists have time and time again documented cancer and other severe health problems in workers who make synthetic textiles.[53] To name but a few: chronic bronchitis and pneumonia, general breathing difficulties and decrease in lung function, sinus infections, and asthma.[54] One 1975 paper described the autopsies of 7 workers in the synthetic textile industry who'd come down with lung ailments, including a 54-year-old woman admitted to the hospital with breathlessness.[55] She died three weeks later. Alveoli in parts of her lungs "were completely obliterated by large sheets of scar tissue," where doctors found polyester fibers. The other 6 workers were similarly afflicted with lung problems, leading the authors of the paper to brand this exposure to synthetic fibers a "new occupational disease." Other studies have linked textile production with higher risk of cancers of the respiratory and digestive systems.[56] (Still more that studied workers exposed to PVC dust have also found higher risk of lung cancer, but not colorectal cancers.) Workers who handle flock—tiny synthetic fibers added to a textile to make it feel velvety—have it so bad that doctors gave it a name: flock worker's lung.[57] One study of 165 workers found that they had a 48-fold increase in the rate of interstitial lung diseases, which cause scarring, that waned after they left the nylon flocking industry.[58]

Lung cancer develops because of mutations: when a cell that makes up the airway divides and replicates, there's a chance that an error in the genetic code will slip in. "Those happen all the time," says Rizzo. "And sometimes those mutations are significant enough that they cause an abnormal growth. If they're caught early by the immune system, it's stopped and the cancer won't develop." But if the mutant is particularly aggressive, the growth develops into a cancer. Carcinogenic substances trigger that disruption of genetic code during cell division. "That's really what a carcinogenic agent is," Rizzo adds. "Tobacco smoke contains a number of chemicals, maybe 70 or 80, that can trigger that."

Plastics, too, are built of carcinogenic chemicals. The monomer vinyl chloride, which becomes PVC, is associated with an increased risk of leukemia and cancers of the brain and lung, according to the National Cancer Institute.[59] Polystyrene is made of benzene and the monomer styrene, both carcinogens. Plastics can also include acrylamide, acrylonitrile, and epichlorohydrin, all probable carcinogens according to the EPA,[60] and are laced with toxic metals that are either known or probable carcinogens, like lead and mercury.[61] As the particles move through soil or air or water, they gather still more "persistent, bioaccumulative, and toxic" chemicals, known as PBTs. "In effect," says the EPA, "plastics are like magnets for PBTs."[62] Heavy metals attach to microplastics with alacrity: one group of researchers found concentrations on particles that were 800 times higher than in their surrounding freshwater environment,[63] and another group found other toxic chemicals on pellets in concentrations a million times higher than in surrounding seawater.[64] Microplastics also pick up high levels of persistent organic pollutants like DDT, a toxic pesticide that the EPA banned in 1972 yet remains in the environment.[65] Because microplastics so easily take to the atmosphere, they could well be picking up poisons in more polluted parts of the world and carrying them to cleaner places thousands of miles away.

A jagged microplastic itself may also scrape up lung tissue, as asbestos does. Particularly tiny particles—down below five micrometers or so—can evade the mucociliary escalator entirely and end up in the alveoli. "That's why we want to avoid any particles of that size getting deep into the lungs, especially if we know they contain toxins that are known to be bad for the lung, like the heavy metals, like formaldehyde, like benzene," says Rizzo. "All those things are known to be carcinogenic and irritating to the airways." And all of those things are found in microplastics. Reaching the alveoli, where oxygen is absorbed into the blood, a particle may leach carcinogenic chemicals that also enter the bloodstream. Particularly tiny particles may *themselves* enter the bloodstream through the alveoli.

Now, of course we have to consider exposure here. Finding asbestos fibers and microfibers in the lungs of people who worked in factories producing the materials is one thing, as they were exposed to very high concentrations, often for decades. Your average human gets nowhere near that exposure. But living in modern times, totally surrounded by plastics, you and I are exposed to airborne microplastics everywhere we go, not just where we work—when we're sitting at home or in our cars, when we're walking down the street, when we're sleeping in our synthetic sheets. We're inhaling microplastics and nanoplastics 24 hours a day, so even people who don't work in textile production are turning up with particles in their lungs. In 2021, researchers analyzed the lung tissue of 20 nonsmokers who'd undergone routine autopsies in Brazil.[66] Only one of the patients had had a history of chronic lung disease, though some causes of death were respiratory-related, like blood clots in the lungs. The scientists discovered a total of 31 synthetic polymer particles in 13 of the 20 people—87.5 percent were fragments and 12.5 percent were fibers. Scaling up the number of particles they were finding in these small bits of autopsied tissue to the weight of a full respiratory system, they calculated that an average person might have 470 microplastics in their lungs. Emphasis on *average*: these patients were salespeople, teachers, and farmers, not textile workers. Yet another 2022 study analyzed tissue samples from 11 people and found significantly more microplastics in the lower regions of the lung than the middle or upper, suggesting that particles are more liable to get stuck the deeper they go.[67]

How much microplastic in the lungs is too much microplastic in the lungs is a question that scientists like respiratory immunologist Barbro Melgert, at the University of Groningen in the Netherlands, are trying to answer. Melgert grows reproductions of airways and air sacs, known as organoids, out of human lung cells, which she then exposes to microplastics without the obvious ethical no-no of forcing people to inhale the particles. "These microplastic fibers, made from either polyester or nylon, specifically inhibited outgrowth of lung tissue," says Melgert. Normally, healthy growth builds the structures of the lungs in children or repairs the lungs of adults with respiratory disease. "Which is sort of worrying because children have a smaller surface area. And if they breathe in the same amount as an adult would, then we would expect negative effects to occur earlier. But also people with lung disease, you don't want to make it harder for them to repair their lungs." Melgert has run these experiments with pristine microfibers and microfibers harvested from store-bought textiles, which are more weathered, and found the effects from the latter to be even more profound.[68] She's also monitored the gene expression within the organoids, a signal of how the lung cells were functioning. Bad news there as well. "I've never seen so many genes changed," Melgert says. More specifically, the microplastics were breaking the mechanism by which healthy lungs repair themselves by mobilizing "progenitor" cells. Normally these would start dividing to patch up the damage in a lung. "We see that these progenitor cells are changed, and then some cells cannot be formed anymore."

Melgert can't yet say what in the plastic is causing this, as there's a long lineup of chemical suspects to wade through. But by creating a leachate from the microfibers—not unlike what Edward Kolodziej and his colleagues at the University of Washington did to find the substance in car tires that was killing salmon—she could apply the chemical cocktail to the organoids, and found the same effect. So it's not just the particle *itself* that's affecting the tissue, but what's leaking out of it. Now it's just a matter of testing the component chemicals on their own, one by one, to narrow down the culprit, or culprits. "We still don't know what is making this change," Melgert says. "It's very frustrating."

Here Melgert's shown that microfibers attack lung tissue, a crucial experimental step toward understanding the mechanisms by which the particles cause respiratory problems, as we've seen in the real world with textile workers. But the average person has a far lower exposure to microfibers, so at what point do the particles start causing issues? Might ubiquitous microplastic pollution have something to do with the growing rates of respiratory disease around the world? "We see asthma growing up, we see COPD [chronic obstructive pulmonary disease] going up, we see lung fibrosis going up, we see lung cancer going up, especially in women," Melgert says. "Then the big question is, Why is this?"

Here we have to tread carefully around a classic scientific trap: correlation does not imply causation. Rates of respiratory disease have been increasing in lockstep with the production and use of plastic. Between 2001 and 2009, the number of Americans afflicted with asthma grew from 20 million to 25 million, according to the Centers for Disease Control and Prevention.[69] Over the same period, global plastics production grew from 480 billion pounds to 635 billion pounds.[70] The two trends *correlate* in their growth rate—almost exactly so—but science doesn't yet have evidence that microplastic pollution, especially in the home, is *causing* more respiratory disease. Scientists have shown, though, that the particles worsen asthma in mice, and even healthy rodents show an inflammatory response.[71] Plus, we know how much household dust is loaded with microplastic now, but we don't have any idea about decades prior, since scientists just began studying it—there's no trend to interrogate, because there's no data.

In addition, particulate matter in general has been linked to cognitive problems—exposure to traffic pollution, for instance, is associated with a higher risk for Alzheimer's disease and other forms of dementia[72]—leading scientists to speculate whether microplastics have anything to do with rising rates of neurodegenerative diseases.[73] That remains speculation, though, because correlation does not imply causation. Air pollution is a wildly complex amalgam of soot, exhaust, pollen, toxic gases like ozone, and microplastics. "There's actually a lot of epidemiological evidence that particulate matter predisposes you to develop lung disease, or to lose lung function, more quickly," says Melgert. "Children who live right next to highways have a lower lung function and a higher risk of developing asthma. So it's not just the person with preexisting disease—the ones that are now healthy have a higher chance of developing the disease."

A baby's first breaths draw in microfibers floating around the delivery room. Then they crawl through microplastic-laden dust, and later bash each other with microfiber pillows. Their lungs are developing all the while, and petroparticles may be getting caught in their airways and evading the mucociliary escalator. Like baby fish, their immune systems also aren't yet fully developed, so children suffer higher rates of respiratory infections than adults.[74] "I think the developing lung is a real concern, because of course if these plastics are not removed when they're children, they will not break down," says Couceiro, who did the study of microplastics in the family home. "We don't know what it's doing to us, but we desperately need to find out. Because if there are—if you like—these hidden impacts that we're associating with asthma, but actually if it is something that's caused by plastic, then we need to know that, so we can start to put interventions in place."

## Chemical Culprits

Of particular concern for children and pregnant women are the many endocrine-disrupting chemicals, or EDCs, in plastics. Your endocrine system is a network of glands that secrete hormones: the pituitary gland and hypothalamus in your brain, the thyroid in your neck, the pancreas in your gut, the ovaries and testes. This network churns out estrogen, testosterone, melatonin, and insulin, among others, which together regulate your metabolism, growth, stress, sleep, and immune system. Without a functioning endocrine system, you can't have a functioning body. Just one mishap somewhere in the system can lead to severe disease or death—consider how critical insulin therapy is to diabetics.

EDCs are a huge family of substances that assault this system, interfering with the body's hormones: of the 350,000 chemicals currently registered for production and use in industry, a thousand may have endocrine-disrupting properties, according to one conservative estimate, and a good many of those are in plastics.[75] Most EDCs have a similar structure to hormones, so they bind to receptors in the endocrine system, preventing our own hormones from doing so or eliciting a response when there shouldn't be one. (Some EDCs, particularly heavy metals like lead and mercury, still interfere with the endocrine system even though their structure is markedly different than hormones.) EDCs also interfere with enzymes the body uses to manufacture hormones, and with enzymes that should then be breaking down those hormones.

Health experts are particularly concerned about three classes of EDCs incorporated into the polymers we're exposed to every day: bisphenols, phthalates, and PFAS.

Bisphenols are most infamously represented by bisphenol A, or BPA, which is a synthetic estrogen but is also good for making hard, clear plastics like water bottles and other food and beverage containers.[76] "Unfortunately, it doesn't just stay in plastics, because as those plastics get older or heated or washed, the plastic breaks down and BPA leaches out," says Jodi Flaws, a reproductive toxicologist at the University of Illinois at Urbana-Champaign, who led a major 2020 report from the Endocrine Society on plastics and EDCs. Between 90 and 99 percent of people worldwide have BPA in their bodies.[77] "We know from many studies now that it can affect brain development and behavior in both animal models and people," says Flaws. "BPA exposure has been linked with anxiety and depression, hyperactivity, and attention and behavioral problems." Same goes for polycystic ovarian syndrome and sexual dysfunction in men. "It's been linked in human studies with breast cancers and prostate cancers, as well as some indication it might be linked with ovarian cancer and endometrial cancer."

The FDA banned the use of BPA in baby bottles and sippy cups in 2012.[78] But a study a few years later that tested 59 baby teethers discovered that they all leached BPA, even though 48 of them were marketed as BPA-free.[79] Bisphenols are also common in both synthetic and natural fibers, and in fact one experiment found much more in wool than polyester.[80] A 2021 investigation by the Center for Environmental Health discovered BPA in concentrations up to 31 times the safe limit in 84 sock brands for adults and infants, including Hanes, Fruit of the Loom, and Adidas. That's especially unsettling given the Pig-Pen effect: the fabric is shedding when you walk around the house, thus contributing to the floor dust that infants crawl around in. That and BPA is easily absorbed through the skin.[81] And the investigation just looked at socks—there's no telling how much of our other clothing is laced with BPA, not to mention the synthetic sheets we sleep in. While everyone was worried about BPA in water bottles—and rightfully so—we've been wrapped in BPA the whole time.

Given the backlash against BPA, plastics producers have been switching it out for other bisphenols, like bisphenol S and bisphenol F, which haven't been as well studied as BPA. The teethers study found BPS and BPF leaching out too. "So you'll see like, *Oh, BPA-free*, but that does not mean it has not been replaced with something very similar in structure to BPA," says Flaws. "We know some of those replacements are as toxic and maybe even *more* toxic, but a lot of them haven't been studied." And replacing BPA with BPS or BPF won't magically eliminate humans' exposure to BPA. Yes, if you're drinking from a BPA-free water bottle you're not drinking the chemical, at least if the manufacturer is being honest with you (and that's a big if). But you're still bombarded with *legacy* BPA: plastics companies began churning out polymers containing the chemical in the 1950s—by the late 1980s, production of BPA in the United States alone reached a billion pounds.[82] All the while, BPA-laced products have been shedding microplastics that blow across the land and swirl in the air and churn in the oceans. Even if production of BPA ceased tomorrow, the endocrine-disrupting chemical would still find its way into humans.

Joining BPA and its ilk in their assault on the endocrine system are phthalates, plasticizers that give the material more flexibility. Plastics can be up to 60 percent phthalates by weight.[83] Accordingly, indoor air is teeming with the chemicals, which researchers have found more in infants than adults.[84] Like bisphenols, phthalates disrupt the reproductive system and have been associated with reduced testosterone and estrogen levels—toxicologists have identified them as possible culprits in declining fertility among men and women.[85] One study of 139 women found that the higher the phthalate levels in their blood, the more likely they were to develop postpartum depression.[86] Studies have also linked phthalates to diabetes and heart problems.

Along with BPA, phthalates are obesogens, in that they increase obesity, and both are classified as metabolism-disrupting chemicals, or MDCs. Laboratory experiments on mice, for instance, show that exposing them to microplastics makes their metabolism go haywire and reshuffles their gut microbiota. And scientists are investigating whether plastics have a role in the obesity pandemic: plotted together since the mid-1970s, the US obesity rate and plastic production increase in lockstep, just like with asthma rates.[87] Obviously there are many other contributors to obesity, like genetics and the proliferation of processed foods—wrapped in EDC-rich plastic, as it happens. But given that the number of obese people worldwide has tripled in just 50 years, which you can't explain away with genetics, scientists are now scrutinizing plastics as a potential contributor to the public health crisis. One team of researchers took common household items like food containers, plastic wrap, and shampoo bottles and extracted 11 different MDCs, including phthalates, which vigorously induced the development of fat cells in the lab.[88] They also tested plastic flooring and found 10 of the 11 MDCs there too—the most of any of the products, actually—and remember Soltani's finding that such a floor is a significant contributor to indoor microplastics. "Given the potency of the extracted mixtures and considering our close and constant contact with plastics," the scientists wrote in a paper, "our results support the idea that plastic chemicals can contribute to an obesogenic environment and, thus, the obesity pandemic."

Beyond messing with metabolism, phthalates are bad news all around. "When you put that all together—something that's associated with obesity, diabetes, and cardiovascular risks, not to mention all the other consequences—you have to take a step back and think, Well, gosh, is this associated with direct mortality?" asks Leonardo Trasande, an environmental health scientist at NYU Langone Health. To answer that question, Trasande analyzed health surveys from over 5,000 Americans who also provided urine samples to be tested for phthalates. He found that among 55- to 64-year-olds, those with higher levels of phthalates used in food packaging were more likely to die of heart disease.[89] "That's independent of diet, physical activity, smoking—all of the usual suspects of causes that you might otherwise argue would explain what we found," says Trasande. Scaling this up to the whole US population, he calculated that phthalate exposure may lead to between 91,000 and 107,000 premature deaths among older Americans each year. "We found this mortality in the *entirety* of populations 20 and older, so we probably were being very conservative in that estimate," says Trasande. "You have to ask yourself at some level, if this is causing so much disease and disability, why is it being used still? And that's because there's a cost to safer alternatives and companies likely profit from the use of these chemicals in plastic."

Phthalates and BPA are similar in that they break down fairly fast, but that doesn't make them any less of a threat to our health. "Even though we're breaking down phthalates or BPA pretty quickly, we're continuously exposed, which is why we continuously have the chemicals in our bodies," says Flaws. Again, eliminating these EDCs won't magically rid our bodies and the world of them. Remember that as a microplastic particle breaks down, it's continuously exposing fresh plastic at its core to whatever medium it's in, or to whatever organism has ingested it—human or otherwise. "These microplastics are everywhere," says Flaws. "And so what's happening is we're being constantly exposed."

The third category of endocrine-disrupting chemical, though, is far more persistent than phthalates or BPA. More than 4,000 kinds of per- and polyfluoroalkyl substances, or PFAS (scientists pronounce it *PEA-fass-iz*—it's an unwieldy plural acronym), are in use, making plastics and clothing water- and stain-resistant. One investigation found PFAS in three-quarters of products labeled as resistant to water and stains, including jackets, shirts, comforters, and tablecloths, all from major retailers.[90] Another study even discovered high levels of the chemicals in anti-fogging sprays and cloths for eyeglasses.[91] Not only are plastics loaded with PFAS, but microplastics gather PFAS already in the environment: one experiment found that particles that'd grown a plastisphere accumulated 85 percent more of the chemicals than pristine particles.[92] "These are actually pretty scary compounds to me because they're what we call 'forever chemicals,'" says Flaws. "They really do not break down very well in the environment." We're talking lifespans of thousands of years.[93] And PFAS are no less harmful than phthalates and bisphenols: toxicologists have shown them to disrupt the immune system and affect the liver and thyroid. They're associated with reduced fertility and low birth weight and have also been linked to breast, ovarian, and testicular cancers, plus non-Hodgkin's lymphoma.

What makes endocrine-disrupting chemicals even more insidious is how their dosing works. The old saying goes that the dose makes the poison: even aspirin is toxic if you eat enough of it. "But what we've learned from many hundreds of studies on endocrine-disrupting chemicals is their dose response is what we call non-monotonic, which basically means nonlinear," says Flaws. So you'll see a strong effect with a low dose, but this wanes with a moderate dose, before the effect rises again with a high dose. If you plotted this on a graph, with a low dose on one end and a high dose on the other, you'd get a U-shape for toxicity, not the linear line that indicates that a higher dose, and a higher dose alone, makes the poison.

You don't need to consume much of an EDC, then, for it to have an effect. Indeed, researchers have found that small amounts of both BPA and BPS cause severe brain damage in goldfish, disrupting the signaling between nerve cells, which they note is highly likely to also happen in humans.[94] (This also amplifies the concern that BPS and other alternatives to BPA may be just as harmful, if not more so.) Toxicologists don't yet fully understand the mechanisms at work here, but at low doses, an EDC could be binding to some receptors in the endocrine system and eliciting the response. "But if the dose gets too high," says Flaws, "you might be shutting off, or down-regulating, those receptors, or shunting it to other receptors where there's an ability to bind and elicit different responses. And in many ways, natural hormones work that way as well." Low doses of estrogen, for example, bind to receptors, but as the dose increases, the receptors degrade until they switch off, a process known as receptor desensitization. Thus you don't get a response at really high levels. "It's not surprising that chemicals that mimic natural hormones could act in a very similar way," says Flaws.

Mothers not only pass the harms of endocrine-disrupting chemicals on to their fetuses but on to even more distant generations. When a mother is exposed to EDCs, so too are her fetus's germ cells, which develop into eggs or sperm. "It's thought that during that exposure, the chemical can target those germ cells and do what we call reprogramming, or making epigenetic changes," says Flaws. "That can be a permanent change that gets carried through generations, because those germ cells will eventually be used to make the next generation, and those fetuses will have abnormal germ cells that would then go on to make the next generation." In the mid-20th century, scientists documented this in women who took a synthetic form of estrogen, called diethylstilbestrol or DES, to prevent miscarriages.[95] The drug worked as intended, and the women gave birth to healthy babies. But once some of those children hit puberty, the girls developed vaginal and breast cancer. The boys developed testicular cancer, and some suffered abnormal development of the penis. Scientists called them DES daughters and sons. "When those DES daughters and sons had children, we now have DES granddaughters and grandsons, and a lot of them have increased risk of those same cancers and reproductive problems," says Flaws. "Even though it was their great-grandmother that took DES—and they don't have any DES in their system—their germ cells have been reprogramming, and they're passing down some of these disease traits."

And now toxicologists are gathering evidence that mothers are passing microplastics and nanoplastics—complete with EDCs and other toxic substances—to their fetuses. In 2021, scientists announced that they'd found microplastics in human placentas for the first time, both on the fetal side and maternal side.[96] Later that year, another team of researchers found the same, and they also tested meconium—a newborn's first feces—and discovered microplastic there too.[97] Children are consuming microplastics, then, before they're even born.

Beyond providing nutrients for the baby, the placenta manufactures a plethora of hormones critical for healthy growth. "It is very important to know what is happening during this early life stage, because we know it's the most sensitive stage of human development," says Hanna Dusza, a developmental toxicologist who's studying microplastics at Utrecht University. "Any perturbation to these development processes can have quite big consequences for later health in an adult life. So it's really, really important to be looking at that as well. And we have indication that indeed microplastics can pass the placenta and expose the developing fetus." In the lab, Dusza grows placental cells and exposes them to microplastics and nanoplastics. She's found that the cells uptake both sizes of particles, but nanoplastics much more so. Other scientists have injected nanoplastics into the tracheas of pregnant rats—it's a more precise way to simulate inhalation because they know exactly how many particles the body's absorbing—and detected the plastics in the fetuses' hearts, lungs, and brains.[98] "I think nanoplastics, we'll find out later on that they're a bigger problem most probably, because they have a bigger surface area," says Dusza. That means more opportunity to leach EDCs and other toxicants.

This is not the time to toy with a human's development, especially with the hormones orchestrating it all. They stimulate fetal growth, initiate the differentiation of cell types—hair, muscle, blood—and control communication between a mother's body and her child's. As the youngster grows, the body passes through "windows of susceptibility," as toxicologists like Dusza call it, in which different hormones play different roles. For instance, there's a period when the limbs are developing, or when there's increased brain development. It's a miraculously well-choreographed ballet that's easily thrown into chaos by EDCs, which scientists have already linked to cancers and lower IQs in children exposed in the womb. Later on in life, too, people who were exposed prenatally to EDCs are more likely to develop breast and ovarian cancer, and the chemicals have been linked to lower sperm quality. "We can already see that there could be a potential problem," says Dusza. "What we don't know yet is actually how many of these chemicals—and what those chemicals are—that are associated with plastic particles. So there's still a big knowledge gap in actually distinguishing the *particle* toxicity and *chemical* toxicity. We know that it's not only the particles that may cause the problem but also different chemicals leaching out of those particles."

The issue is—again—that plastics manufacturers don't disclose the ingredients in their products, which leaves the job to chemists to reverse-engineer the recipes. According to one estimate from researchers at ETH Zürich, at least 10,500 substances are used in plastics, and probably more in actual fact. (That study on MDCs in household plastics identified the signatures of 55,000 unidentified chemicals.) Of these, they identified over 2,400 as "substances of potential concern," meaning they're either persistent (like PFAS), bioaccumulative (the body can't easily break them down), carcinogenic, or just straight-up toxic.[99] Half of these substances of potential concern aren't regulated in the US, Europe, or Japan, but 900 of them are approved in those places for use in food-contact plastics. "It's a huge variety of substances," says environmental chemist Zhanyun Wang, who did the tally. They include our usual suspects like heavy metals but also chemicals that are meant to replace toxic components of plastic but are themselves toxic. "This regrettable substitution is also happening," says Wang. "Unfortunately, our current chemical regulatory system does not really require the manufacturer to say that, *OK, we're replacing with this chemical, and this is the reason why.*"

And the physical structure of microplastics can turn nonpersistent chemicals persistent. Phthalates, for example, degrade far faster than PFAS. But microplastics pack these chemicals in their core, so they aren't initially exposed to the environment. "They can be a shield for more easily degradable chemicals," says Wang. As the microplastic breaks apart, it sheds the toxicants at its core, releasing them into the surrounding medium, be it seawater or the human body. So even if a chemical isn't persistent on its own, packaged in a microplastic, it's offered protection—a sort of slow-release poison pill.

## Plastics in the Gut, Blood, and Brain

It's a pill we'd rather not swallow. Analyses of human stool have turned up plenty of microplastics that pass through us, but the particles also seem to be getting stuck in the intestines. In one study, 11 patients in Malaysia who'd undergone colectomies—the removal of portions of the colon—provided tissue samples.[100] Nine of the patients had colon cancer and two did not. Researchers sifted through the samples and found an average of 800 microplastics per ounce of tissue. They noted the prevalence of transparent filaments, suggesting the digestive system bleached the particles, an indication that the colorants and other additives in microplastics leach out in this hot and acidic environment, though this may instead be due to the fact that single-use plastic food wrappers tend to be clear, so the patients could have eaten already-transparent microplastics. But in other lab experiments, scientists have shown both how microplastics greedily gather metals and then release them in a solution that mimics the chemistry of the human gut—the smaller the particle, the stronger this Trojan horse effect.[101] It's worth noting again that textile workers have increased rates of lung cancer *and* cancers of the digestive system—the mucociliary escalator brings particles out of the lungs to be swallowed.

Microplastics and their component chemicals might also pass through the gastrointestinal wall and into other tissues, the kind of translocation we've seen in fish. Our guts are lined with special cells that soak up nutrients and pass them into the bloodstream, and immunologist Joost Smit of Utrecht University has shown that in mice at least, microplastics go along for the ride. He fed particles to the animals and then tested their blood, and the microplastics were there 10 minutes after they'd entered the rodent's digestive system. Smit also found the particles in the livers of the mice. "It's very low amounts of plastics, which can travel throughout the body after dosing, but they are there definitely," Smit says. "Especially the smaller microplastics, but also the bigger microplastics, you can see them in the blood after you dose animals. And even the 10-micrometer ones, which are quite large microplastics, they're the size of half a cell basically, they still are transferred over the intestine."

Sure enough, in 2022 scientists reported finding plastic particles in human blood samples for the first time.[102] Whether the pollutants had been absorbed through the gut or alveoli in the lungs wasn't clear. Either way, by entering our bloodstreams and flowing to our organs, microplastics may Trojan-horse whatever they'd accumulated while wandering around the environment. "They could be a kind of a vehicle of possibly dangerous substances that then get transferred in our body," says Smit. "I think that's the highest risk of these particles. Themselves they don't seem to do much. But the stuff they carry, that might be a potential big risk." Obviously our guts aren't what you'd call a sterile environment—they're teeming with all manner of bacteria and viruses, and indeed that microbial messiness helps our bodies absorb nutrients in the first place. "But the one thing which is kind of sterile is our blood in the body itself," says Smit. "And of course, when a particle is loaded with bacteria, and it can cross the intestine and reach the bloodstream, it's something which might trigger the immune system."

Like the particles may act as nuclei in the atmosphere, collecting water and ice, so too might they promote clotting in the bloodstream.[103] "Obviously, your blood clotting in your body when you're not bleeding is bad," says Nathan Alves, a chemical and biomolecular engineer at the Indiana University School of Medicine who's studying microplastics. Blood clots in the lungs and brain are particularly dangerous, but clots in the legs can also dislodge and flow to the lungs, where they cause pulmonary embolisms, which also kill. "When you stick something of reasonable size in the blood that starts attracting things, then you tend to aggregate. And once aggregation starts, it tends to be hard to stop until the blood stops flowing." It's a runaway train—made of blood. Around 8 percent of the population has some sort of variation in their blood that makes them more susceptible to coagulation. "I think in those kind of cases, when you add in an additional perturbation, like microplastic, you're seeding more coagulation," says Alves. "Maybe this is actually an underlying thing that's happening more commonly clinically than we realize, but we're not looking for it." A big enough blockage somewhere in the body can make the heart overwork and fail. Or the heart will enlarge and maybe you survive, but now you have cardiac damage. "The other side of things is if you get a hugely massive exposure in your lungs to something, then the oxygen diffusion across the alveoli will drop," says Alves. "And then you have less oxygenated blood coming into your body, and muscles like your heart might actually start to starve a bit. And you could cause cardiac damage."

But back to the microorganisms on microplastics, because we've got fungi to worry about too. In Kenya, scientists discovered a horde of fungal species growing and reproducing on soil microplastics, the core community of the plastisphere consisting of pathogens that infect plants, animals, and humans.[104] "We found that certain pathogenic groups are concentrated on plastic, meaning that they're in a higher concentration in the plastisphere compared to the soil," says University of Bayreuth microbiologist Gerasimos Gkoutselis, who led the research. The most dominant fungi on the particles were nasties like *Naganishia albida*, a yeast that causes pneumonia,[105] and *Rhodotorula*, which is notorious for colonizing plastic equipment in hospitals, like catheters. "They really use plastic as an entrance point into the human body, for example during surgery," says Gkoutselis. Once inside, *Rhodotorula* triggers serious infections.

The concern isn't just that microplastics could act as vehicles for delivering such pathogens into the human body but that they could be delivering them around the world. When winds loft microplastics into the atmosphere and carry them thousands of miles, that might transport pathogens "across boundaries, across biomes, and therefore transports the fungi to new habitats and enlarges their range," says Gkoutselis. Many of these fungal pathogens are "opportunistic," meaning that in the right conditions, a single species can infect either a plant or an animal, including humans. Microplastics could spread such nondiscriminatory menaces where they don't belong.

To be clear, while we know microplastics are in our blood, scientists haven't yet found them to transport anything dangerous from the outside world into our guts and then into the rest of our bodies. But Smit's work shows that mice guts do absorb microplastic particles, and our guts probably do the same. And remember the extraordinarily high doses that infants get in their formula—a daily synthetic stew of millions of microplastics and trillions of nanoplastics that may be crossing the gastrointestinal barrier and moving through their developing bodies.

Even if microplastics just stuck around the digestive system, they might mess with our microbiomes both by introducing foreign species and by acting as platforms on which bacteria grow. Experiments with a machine that simulates a full human digestive system show that PET particles—in realistic concentrations we're actually eating—had an antibacterial effect, decreasing the number of viable bacteria and thus compromising the microbiome.[106] These bacterial species not only promote healthy digestion but also help maintain the immune function of the intestines and reduce inflammation by producing anti-inflammatory molecules. Intriguingly, a separate study found that subjects with inflammatory bowel disease, which may be caused by a malfunctioning immune system, had 50 percent more microplastics in their stool than healthy people.[107] That's not to say this proves microplastics *cause* IBD, though, as the disease may somehow make the gut retain more of the particles for scientists to find.[108] But if microplastics are knocking out the bacteria that promote a healthy digestive system, that could be a causal link that demands more research.

A battery of other questions remains here. Might other gastrointestinal issues, like celiac disease and lactose intolerance, change how microplastics interact with the digestive system? Do the fat and fiber and sugar content of foods influence how the particles leach chemicals? And are microplastics staying in our bodies long enough to fragment into nanoplastics, as they do in the guts of krill and earthworms? I'm not saying that our digestive systems are anything like that of a worm or planktonic crustacean, but if we're also breaking down the particles, that'd mean microplastics that are too big to be absorbed through the gut degrade inside us into nanoplastics that can be.

What these nanoplastics might do to the human body, scientists are just beginning to investigate, but the implications aren't great. In the remote Alps, millions of nanoplastics will fall on your head, but in your home, you're surrounded by plastic in couches and carpets and sheets, all of them spewing particles that accumulate in dust and get kicked back into the air for you to breathe. And the smaller a particle gets, the more places it can squeeze into, physiologically speaking. We might even absorb them through our skin, particularly via hair follicles—*that's* how small these things are. And that's also why we need more research on whether the loads of plastics in cosmetics could be entering the body.

Scientists already have evidence of plastics readily moving through the human body.[109] Autopsies of 29 patients who'd received hip or knee replacements, made of metal and polyethylene, turned up traces of both materials in their lymph nodes, livers, and spleens.[110] Though the doctors who wrote up the findings noted that in the majority of the patients "the concentration of wear particles in these organs was relatively low and without apparent pathological importance," the prevalence of particles was greater in patients whose implants had failed. In one of these cases, small lumps of immune cells known as granulomas, which form due to infection or the presence of foreign objects, popped up in the liver, spleen, and lymph nodes "in response to heavy accumulation of wear debris from a hip prosthesis with mechanical failure." The patient also had compromised liver function.

And then there's the brain. Like fish, we have a blood-brain barrier that's supposed to keep foreign objects out of the organ, but the littlest of particles can slip through. Scientists have long studied the effects of metal and metal-oxide nanoparticles—tiny bits of silver or gold or copper oxide—on the human body. "Several studies demonstrated that these particles can be taken up by our bodies and can actually reach the brain," says Remco H. S. Westerink, a neurotoxicologist at Utrecht University.[111] "Some of the particles can even damage the protective blood-brain barrier, thereby facilitating their own entry as well as the entry of other unwanted chemicals. Notably, decreased blood-brain barrier function and neuroinflammation are also hallmarks of most neurodegenerative diseases." Metal nanoparticles can even get taken up by nerve endings in the nose, thus bypassing the blood-brain barrier entirely. "Once in the brain," Westerink adds, "these particles can have all kinds of effects, but the most predominant and adverse effect is the generation of oxidative stress, which damages the brain cells and can even kill them." In severe cases, small parts of the brain are lost, and the organ begins to age faster, thereby speeding up degenerative diseases like Parkinson's. But the effect depends on what the nanoparticles are made of: iron oxide, silver, and copper oxide are potent, whereas gold and titanium dioxide are more inert and cause less damage. "Nevertheless, if the exposure is sufficiently high, even these inert particles can still induce damage," Westerink says. "So in the end, whether or not the exposure causes adverse health effects is determined by the degree of exposure."

Do nanoplastics behave like metal nanoparticles in the human body, suffusing through our tissues and entering the brain? Another team of researchers confirmed that, indeed, mice fed fluorescent plastic particles smaller than two micrometers ended up with the foreign objects in their brains, so the stuff passed through the gut wall, into the blood, and then across the blood-brain barrier.[112] Microglial cells—which remove damaged neurons and infections—recognized the particles as a threat and attacked them, but this significantly impacted the cells' ability to proliferate, and in the end they perished. That means plastic particles are neurotoxic, at least in the brains of mice, but given that we're also mammals, there's no reason to believe that little bits of plastic can't do the same in our bodies. Indeed, in several other experiments, toxicologists have found that microplastics—in concentrations we're already inhaling and eating—kill human cells grown in the lab.[113] "Our preliminary data also indicates that plastic nanoparticles, and even microparticles, can reach the brain," says Westerink. "While there is only very few data on plastics in the human brain, I don't have much doubt that these particles can reach it. Whether or not the particles cause harm to the brain is a different question."

That, again, comes down to exposure: are we eating and drinking and inhaling enough nanoplastics to have an effect? Only a fraction of metal and metal-oxide nanoparticles make it into the brain, and scientists are using high concentrations in lab experiments. But for practical purposes, the experimental exposure is limited to hours, days, or months. "In real life, however, exposure is at lower levels, but it is continuous over an entire lifetime," Westerink says. Your whole life you've been taking in microplastics and nanoplastics, and day by day, you're exposed to more and more plastic as production accelerates and Earth's natural systems grow increasingly plasticized. When, then, does the exposure become unsafe, if ever?

We can say with certainty that our brains would much rather avoid nanoplastics. Unlike a metal nanoparticle, a nanoplastic isn't a single element, but a suite of monomers, plasticizers, flame retardants, and on and on. "To make it worse, plastic particles undergo aging and weathering in the environment, which may affect their properties, and they can even adsorb different environmental chemicals," says Westerink. "This results in literally an endless variation of different types of plastic particles." Hell, a plastic particle that enters your body may have previously traveled through the digestive system of some sea critter.

For the time being, though, scientists don't *yet* have evidence that microplastics and nanoplastics—at current concentrations in the environment and in homes—harm human health. The stuff isn't good for us, that's for sure, and EDCs on their own are proven poisons of the highest order. But thus far the only clear harm is in textile workers and people exposed to PVC dust, who are in contact with way more particles than you and me, though finding microplastics in the malignant lung tissue of non–textile workers isn't a good sign. Microplastics are in our bodies, that's also for sure, but what they might be up to remains a big, frustrating unknown.[114] "The effects of microplastics on human health, we maybe know 1 percent of it," says Smit. "We still don't know 99 percent of what they might do, or actually what they might *not* do."

We humans are more sullied than any other animal on Earth. We spend our whole lives in contact with polymers, breathing and eating and drinking plastics, whereas a fish has to deal with what's diffused into the environment. Which is a lot, but a fish doesn't spend 90 percent of its life indoors huffing bits of clothes and carpets and couches. "I'm absolutely certain that using the right tools, we will find it in every part of the human anatomy, and that means it's in every part of every animal on the planet," says Steve Allen. "Now, as the levels go up, that's the part that we don't know about. Just how far does it have to go before we start having ecosystem collapse? I think that's probably on the cards."

CHAPTER 5

# Turning Down the Plastic Tap

On September 8, 2018, a blue ship sailed under blue skies, towing what we can safely call the strangest cargo ever to pass through San Francisco's Golden Gate: a 2,000-foot-long plastic tube meant to scrub the ocean of . . . plastic.[1] Like a tadpole with a great black tail, the vessel chugged along as no-doubt-confused tourists watched from the famous bridge above. They weren't the only ones, as it happened, because oceanographers were also confused as to why this, of all things, was the ultimate weapon with which humanity would fight plastic pollution.

The idea was to tow the device out to the Great Pacific Garbage Patch, form it into a U shape, wait for all the plastic to float in, and periodically send a boat to scoop out the catch. In actual practice—and this is what the oceanographers were warning about—the sea would have other plans, like trying to snap the thing in half. And indeed just a few months after it launched the contraption, the Ocean Cleanup nonprofit—which had raised $40 million for the project—announced that the giant tube was now in two pieces, and they were towing it to Hawaii for repairs.[2]

Forget for a moment the moxie of trying to engineer a 2,000-foot-long plastic cylinder to survive the high seas without *becoming* plastic pollution. It was the underlying principles that exasperated oceanographers. According to one survey of 15 experts on ocean plastic pollution, more than half had serious concerns about the project, and a quarter thought it was "a bad idea with little or no redeeming value."[3] The central problem was the project operated in two dimensions, while oceanic plastics operate in three: what floats at the surface in the Great Pacific Garbage Patch is a minuscule fraction of what's out there, the oceanographers insisted. If humanity is going to truly address plastic pollution, we have to go farther upstream, keeping microplastics and the macroplastics that spawn them from ever getting to the environment in the first place. Spending tens of millions of dollars on giant oceanic plastic catchers is like trying to drain a tub with the faucet still on, or mopping a floor with floodwater pouring in, or raking leaves at the start of autumn—without stopping the torrential flow of plastics into the environment, we don't stand a chance of fixing the problem. Get far enough upstream, though, and you might actually hold corporations responsible for this planetary vandalism.

At its core, the microplastic crisis *is* the macroplastic crisis. If we can keep a cup out of the environment, we can keep it from exploding into particles—any waste that you see floating in a river or tumbling across the land is just pre-microplastic. "The cup is still there, it's just in a million pieces," says Marcus Eriksen of the 5 Gyres Institute. "At that point, it's impossible to extract from the environment." If we can keep a sweater from shedding microfibers, and a tire from disintegrating, and polymer-based paint from chipping, we can mitigate microplastics. The requisite change is colossal, but not impossible. Plastics aren't going anywhere—they're just too useful and too omnipresent. And even if a virus killed every human next week, our plastic would still decay and flush out to sea and take to the air, until one day a long time from now it will all have decomposed as far as it can go, wrapping the planet in a perpetual nanoplastic haze. But there are ways to at least thin that haze by slowing the emission of plastics of all sizes.

The peril here, though, is falling into the same personal-responsibility trap that the fossil fuel industry suckered us into with climate change and recycling. If only you'd stop flying so much you'd reduce your carbon footprint—BP popularized the term in 2004 with a $100-million-a-year ad campaign—and help save the world.[4] Plastics producers did the same with recycling, hailing it as the cure for pollution when they knew the economics of it were busted (more on that in a moment)—hence to this day barely any plastic is getting recycled. As Larry Thomas, former president of the Society of the Plastics Industry (now called the Plastics Industry Association), told NPR in 2020: "If the public thinks that recycling is working, then they are not going to be as concerned about the environment."[5] Like opioid makers blaming patients for getting addicted to their pills, the plastics industry keeps the world hooked on polymers it can't control, as waste continues to escape unchecked.

So by all means, get a filter for your washing machine—every little bit helps. But we can't lose sight of the underlying systemic problems. Thinking bigger, we need laws in every country mandating microfiber filters on every new washing machine, and we need ways to safely dispose of those fibers. We can't rely on wastewater treatment plants to filter out the particles, because the fibers they do catch just go out with the sludge. "That's where I see the most difficult thing with microplastics, is we probably won't ever be able to treat our way out of the problem," says Scott Coffin, the research scientist at the California Water Boards. "Just because what we're seeing with application of biosolids to agriculture is that eventually it has an impact on the plants. So it's really hard to give recommendations to a wastewater treatment plant operator at this point, because they're kind of stuck between a rock and a hard place."

Yes, engineers can develop filtration systems that snag microplastics in these facilities, the last stop before the open ocean.[6] Some treatment plants already use ultra-fine membranes to recycle wastewater into drinking water—so fine, in fact, that only water molecules can get through, leaving behind contaminants like bacteria and viruses.[7] But in the grand scheme of wastewater treatment, recycling into drinking water is rare because it's so expensive: a facility costs billions of dollars to build and is costly to operate on account of needing a lot of energy to force water through the membranes. Wastewater gets nowhere near this treatment before being pumped out to sea since it doesn't need to meet the same purity standards as drinking water. And even if we had an economical way to filter microplastics out of wastewater, we'd need to deploy it all over the planet—the US alone has over 16,000 treatment plants.[8] And remember that half the wastewater humans produce isn't treated at all before being flushed into the environment.

So we have to move farther upstream still and banish obsolescence. The fashion industry pioneered the economics of obsolescence in the mid-20th century—the easiest way to sell more clothes is to force them out of style, making them embarrassing to wear a short while after buying the things, only for them to come back in style a few decades later, when people buy the same clothes they once threw out.[9] That very profitable innovation has since spread across the consumerist economy, from phones to cars to home decor, producing heaps of wasted, uncool goods. Fast fashion sprinkles in yet more obsolescence by flooding the market with cheap clothes you have to replace long before they go out of style: if your sweatshirt is falling apart after a few weeks, you've already flushed a good portion of it away as microfibers. But keep in mind that natural fibers these days are soaked in additives, so they're not a perfect alternative. Even organic cotton just means it was *grown* without synthetic chemicals—it was still dyed and potentially further treated to be tougher and waterproof.[10] Whatever the material, fast fashion is garbage, literally: 86 million pounds of discarded clothing piles up each year in dumps in Chile's Atacama Desert, where wind and UV bombardment slough off microfibers to blow around the landscape and into the atmosphere.[11] Over in Accra, the capital of Ghana, 40 percent of the 15 million used garments that arrive every week are too busted to resell, so they meet the same fate.[12]

As a consumer, the most impactful thing you can do is buy better clothes and wear them for years—good fashion instincts be damned—since experiments have shown that textiles emit fewer and fewer microfibers over time. But then again, *you* didn't make this mess. "We can educate consumers, but in the end, the responsibility has to be on industry and also obviously on policymakers and regulations," says Laura Díaz Sánchez, who specializes in microfibers at the Plastic Soup Foundation, which advocates for action on plastic pollution. "We know enough to be concerned and to make the fashion industry accountable that they need to make better clothes. Whoever is putting the product out there—any plastic product, not just clothes—should be responsible for what is happening to it after it's been sold. And that's not what's happening now."

Because there are so many pathways for mismanaged plastic to get into the environment—bottles chucked into rivers, microplastics flying off tires, bags blowing off landfills—more recycling alone can't fix this. But it is vital to mitigating the problem. It's just that we have to fix the broken economics first. The reason so little plastic has been recycled isn't that it's impossible to do so, but that it isn't *profitable* to do so. For a waste management company the calculus is simple: can I make more money selling recycled plastic than it took collecting and processing the waste? For the vast majority of the single-use plastic humanity has churned out—wisps of wrapping, little condiment packages, polystyrene foam containers—the numbers just don't add up, so they're diverted to the landfill or incinerated. Simple plastic bottles are easy enough to recycle, but potato chip bags aren't easily processed because they're multilayered materials. That and the more a plastic is recycled, the more its quality degrades—a bottle can't become new bottles in perpetuity. "A pile of gold is always going to be recycled because the value of the gold always outweighs the cost of collecting and processing," says Tom Szaky, founder and CEO of TerraCycle, a recycling company. "So you could litter gold and no one would have a problem. Someone would pick it up and sell it or melt it very fast, and thank you probably for it." By contrast, if you need to empty a portable toilet, you have to pay someone to deal with it.

Over the past decade, three converging trends have made plastic look less and less like gold and more and more like crap, borking the economics of recycling even further. For one, waste management companies in the US have historically plucked out the most profitable items like bottles and simple containers and shipped the rest abroad, particularly to Asia, where labor for sorting the stuff is cheaper—we're talking 2 billion pounds of exported plastic waste each year.[13] China used to import half the world's recyclable plastic but in 2018 closed its borders to polymers.[14] So now even more plastic gets chucked in the landfill, where it breaks into microplastics that contaminate surrounding soils and seep into groundwater.[15] In low-income countries now overflowing with our waste—after China's ban, US exports to Malaysia, Thailand, and Vietnam jumped 300 percent—plastic that can't be profitably recycled is burned in open pits.[16] (This is a perverse inversion of colonialism: instead of violently extracting a desirable resource from other countries, developed nations are now violently injecting an undesirable resource into other countries.)

The second factor is that the price of oil remains low, meaning it's cheaper to just churn out virgin plastic instead of bothering with recycling. This is a rather predictable capitalist disgrace: the all-powerful market is totally fine making the mess, but totally uninterested in fixing it. And third, in their incessant quest for profit, companies have opted for lighter yet more complex packaging. A glass bottle of baby food, for example, now comes in a multilayered plastic pouch, which saves on shipping costs but makes the container a nightmare to recycle. "There's no gold anymore—it's all crap," says Szaky. "Recycling in general is always going to be an imperfect Band-Aid, and I say this as a recycling business. It's critically important, but it's not the silver bullet. There's only one silver bullet, which is buying less."

So how about this: slap a tax on plastic.[17] That'd make it more expensive for producers to churn out virgin plastics, incentivizing companies to switch to more easily recyclable materials like cardboard and glass, or even compostable packaging like banana leaves—supermarkets in Asia already do this—or move away from single-use altogether. (Plastics apologists argue that the material makes goods lighter, which means it takes less energy to ship things. Sure enough. But this glosses over the fact that developed economies use those products and then ship their plastic waste across oceans to other countries, where the material is burned and its component carbon sent straight into the atmosphere. That and there's the loads of emissions from the production of those plastics in the first place.) California did the math and estimated that a tax of one cent per plastic item would net a few billion dollars a year, which could then fund recycling programs and microplastic mitigation projects.[18] So plastic producers, not the people, would pick up the tab for solutions, though you'd have to ensure that they didn't pass the cost on to consumers. But if you're in the market staring at naked produce and produce wrapped in unnecessary plastic, you can vote with your wallet. It's a sin tax, like the tobacco tax, only instead of incentivizing people to not destroy their lungs, you're incentivizing humanity to not destroy the planet. Or countries could mandate by law that plastics producers spend their own plentiful cash on trash wheels and rain gardens and massive recycling programs, finally shifting the responsibility for planetary plasticization from consumers to the corporations churning the pollutant out.

But the need for more recycling may clash with the need for better plastics. As John Boland discovered in his study on kettles, a simple layer of built-up copper oxide stops the release of microplastics. If materials scientists can replicate this so all kettles and disposable coffee cups and baby bottles don't shed particles from their very first use, that'd go a long way in reducing our exposure. "There may be coating solutions that will prevent the release of microplastics—if you can demonstrate that those coating solutions are still recyclable," Boland says. "If you think you have a solution, you need to make sure that the solution isn't just another kind of different problem."

You might question why as a society we would invest so much money and energy confronting the microplastics crisis when the climate crisis is an immediate threat to life on Earth. And to that I'd say the two crises are one and the same. Remember that by 2030, the manufacture and use of plastic is projected to generate as much greenhouse gas per year as 295 coal plants, and that will more than double by 2050 as the industry massively scales up production. And every bit of microplastic out in the environment belches greenhouse gases as it ages. Even after we've decarbonized our civilization, switching to electric vehicles and renewable energy, we'll still in a sense be burning fossil fuels: we're pulling carbon out of the ground, turning it into plastic, and releasing the material into the environment to off-gas that carbon. Worse yet, waste managers staring at ever-bigger piles of plastic are just burning more of the stuff. We'll have decarbonized energy production and transportation, sure, but not the material central to every aspect of our lives. "When you think of a future that has successfully dealt with the plastics crisis, or you think of a future that has successfully dealt with the climate crisis, those futures look pretty similar," says Steven Feit, a senior attorney at the Center for International Environmental Law.

Every way to fight microplastic pollution *is* fighting climate change. Bolster public transportation and it'll take cars—which emit both carbon and tire microplastics—off the road. Switching to electric cars is of course hugely important for combating climate change, but the most impactful thing we can do is getting as many tires off the road as possible, especially since the extra battery weight and torque of EVs could produce more tire particles. Regarding microfibers, if we stigmatize fast fashion and pressure companies to design better fabrics, we'll all get longer-lasting clothes—which take energy to manufacture—that shed fewer fibers. Wash and dry clothes less to make them last even longer, and we'll use less energy in the process. Use less plastic in general and we'll reduce the emissions from their production.

Shifting toward bio-based plastics made from plants like corn and sugarcane—which currently account for 1 percent of global plastic production—would reduce the emissions associated with the extraction of fossil fuels for traditional plastics.[19] But really it'd shift those emissions elsewhere. Growing the plants is an intensive industrialized process that releases carbon, and tilling the land releases carbon stored in the soil.[20] That and you'd need a whole lot of land and water: one estimate reckoned that replacing all the plastic packaging in the European Union with bioplastics would require more cropland than the area of Ireland and a fifth of the EU's freshwater withdrawal.[21] If you wanted to replace plastic packaging globally, you'd need more land than the area of France and at least 102 trillion gallons of water—on a planet where a growing human population needs ever more land and water to grow food. (Yet another reason why it's not as simple as everyone switching from synthetic to cotton textiles, as it would take an astonishing amount of land and water to meet demand.) That doesn't bode well for the environment, not to mention that scientists are still researching the toxicity of bioplastics, which are loaded with the same additives as traditional plastics. One upside to bio-based plastics, though, is that they're made of plants that absorbed $CO_2$ as they photosynthesized, meaning the polymer is made of carbon sequestered from the atmosphere, in contrast to fossil fuel–based plastics, which are made of carbon that's supposed to be locked underground. In that sense, the wide-scale production of bio-based polymers could indeed reduce emissions.[22] But it takes energy to manufacture plastic pellets, ship them around, form them into bottles and bags, and ship those products to consumers.[23] And when bio-based plastics escape into the environment, they shatter into microplastics and off-gas their carbon all the same.

So newfangled materials aren't going to solve the microplastics crisis: the farthest upstream we can go is majorly curtailing the production of plastics. That's how we fix this crisis—full stop. I'm in no way calling for the abolishment of plastics, which would be quixotic, but I am pleading for sanity. Seeing cucumbers wrapped in single-use plastic in the market shouldn't give us peace of mind, but should make us question why produce with perfectly good skins needs additional synthetic skins. So long as we're churning out single-use plastic—doesn't matter if it's made from fossil fuels or plants—we're trying to drain the tub without turning off the tap. We need rain gardens and trash wheels and washing machine filters and better fabrics, and we need them now, but more important than anything else, we've got to cut it out with all the plastic. "Everything that's farther downstream from that is, depending on where you are, either a nice-to-have or a must-have," says Deonie Allen. "But it's the backup—it's not what you're depending on to manage the pollution."

There's precedent here for change. Governments are already taxing or banning bags and other single-use plastics. Smoking used to be the norm, and now it's taboo. And lead used to be in all kinds of products until advocacy groups started suing. "First we sued Squibb, and Johnson and Johnson, and Pfizer, and we got the lead out of baby products," says Michael Green, CEO of the Center for Environmental Health, which did the study on BPA in socks. "Then we sued the toy makers. Then we sued the candy makers, then we sued the baby bed manufacturers, then we sued the bounce house manufacturers, and then we sued the children's jewelry manufacturers. It took us about a decade, and eventually we couldn't find any more lead in any more children's products. Then we went to Capitol Hill with our friends and got legislation passed banning lead in children's products."

The complicated nature of polymers, though, creates a much bigger challenge. Not only do plastics contain lead, they contain a plethora of other toxicants as well. When governments started banning BPA, manufacturers just substituted other bisphenols, and clothing is still loaded with BPA and PFAS because few people thought to *look* in clothing, not realizing it's made of plastic. "One of our demands is don't talk to me about PFOA or PFOS—I want to talk about the *class* of chemicals," says Green. The problem with clothing is that a brand gets textiles from a supplier, often not knowing what kinds of chemicals their customers will be exposed to after inhaling microfibers. "We want them to know what's in their products and we want them to not avoid the bad, but proactively look for the good, to do R&D about how can we make socks that don't fall off the baby's feet and aren't also potentially causing that baby to be obese when they're seven."

This is a matter of equity as well. By 2050, the plastic in the sea will outweigh all the fish, but in many countries overwhelmed by imported waste from the developed world, it *already* outweighs all the fish: subsistence fishers in island nations like the Philippines are reporting catching more bottles and bags than seafood.[24] All that macroplastic is baking in the fierce tropical sun, fracturing into microplastics and nanoplastics that corrupt coastal waters and the fish that swim in them. On land, incinerating plastic in open pits bathes communities in microplastic-laden smoke, an escalating public health crisis in the developing world.

In the US and other developed countries, the disproportionate number of poor folk and people of color living in food deserts—neighborhoods where grocery chains don't bother setting up shop—might only have easy access to processed foods mummified in single-use plastic, maybe from a corner store. And remember what Leonardo Trasande found in his study, that people with higher levels of phthalates used in food packaging are more likely to die of heart disease. "I think about food deserts and the lack of accessibility to plastic-free food, and what kind of impact that might mean for these models that are predicting how much we're eating and we're interacting with," says Imari Walker, the environmental engineer we met in the first chapter. "They might be missing the story, that there are certain communities that might have a much larger interaction with plastic and microplastics, just because of sheer packaging in these industries that they're surrounded by." Same as it ever was: the disadvantaged get exposed to more of everything terrible—heavy metals, air pollution, and now microplastics.

And consider the factor of age here. If you were born in the '50s or '60s, you grew up in a world that was only flirting with plastics. But I'm in my late 30s, so I've lived my whole life bombarded by single-use plastics. Babies today are drinking millions of microplastics and trillions of nanoplastics in their formula daily—quite the christening into the modern plasticine existence. So an 80-year-old living today will have eaten and inhaled far fewer particles than an 80-year-old living half a century from now. Like the environment will only grow more corrupted with microplastics as humanity expels them unabated, so too will our bodies.

Psychologists may question the healthiness of this, but the most impactful thing you can do right now is to get angry. Long pitched as benign wundermaterials, plastics are ubiquitous poisons that have seeped into the roots of the tree of life. Changing your personal habits is all well and good, but it was never going to solve climate change, and it's never going to solve the microplastics crisis. Above all else, we have to elect politicians who understand that fighting climate change and fighting plastic pollution are two sides of the same coin and that increasing exposure to polymers is a public health emergency. "Look, I'm a former federal regulator, so my approach is: adopt effective laws and then strictly enforce them," says Judith Enck, the president of Beyond Plastics and previously an EPA regional administrator, who we met in the first chapter. "I don't think individual consumer action—while laudable—is enough to change the trajectory. This is a systemic problem."

If we *as a civilization* don't stop releasing so many greenhouse gas emissions and so many microplastics, your efforts will be for naught. What's far more valuable is your outrage at the corporations that have plasticized this planet in the sociopathic pursuit of profit, at the single-use mentality that addicted humanity to long chains of carbon, at the tiny toxicant that's corrupted our bodies. Think of the baby fish filling up their bellies with microplastic, of your indoor air swirling with particles, of the shameful fossil record we're leaving behind. Nowhere is untouched, and the crisis will only deepen as microplastics flow unabated into every corner of Earth and, by their very design, persist. "Are we going to go the way of climate change, and not listen to the scientists until it's too late?" Enck asks. "On the plastic pollution issue, I think we still have time."

Geologists have been arguing about this notion of the Anthropocene, a new age in Earth history characterized by humanity's transformation of the planet.[25] The controversy isn't about whether the age exists, but what signal in that geological record should mark its inception. You could make the case that it was the invention of agriculture, when we gave up the freedom of the hunter-gatherer lifestyle for the prison of the farm: body-breaking labor to produce grains that made us sicker yet created food surpluses that fed a booming population[26]—such a big population, in fact, that we had to invent plasticulture to sustain it. You might say it was the Industrial Revolution, when we pulled fossil fuels from the ground and burned them to feed insatiable economic growth. But others argue that the appearance of plastic in sediments should mark the birth of the Anthropocene, when humanity began spewing a noxious glitter that now swirls in the oceans and infiltrates soils and blows around the atmosphere.[27]

Whatever your preference, each option marks a monumental advance in the story of humanity, but each is in fact a progress trap—ingenuity that spawns unforeseen consequences that we either can't fix or don't feel like fixing.[28] Without agriculture we'd be mere pockets of people getting offed by predators, but we'd also be living in harmony with the land. Without fossil fuels we'd still be filling up horses with hay instead of cars with gasoline, but we also wouldn't have to worry about runaway climate change. Without plastics you'd get nowhere near the medical care you do today, but you also wouldn't be eating, drinking, and breathing petroparticles. Agriculture gave way to industrialization, which gave way to plastic, each progress trap digging us deeper into ecological debt. We've come too far as a civilization to admit that maybe, just maybe, our idea of progress is the planet's idea of sweltering, plasticky hell.

Now we have to ask ourselves: is humanity's leg fully caught in the plastic trap, or can we tear it free and limp away? One path is capitalism as usual, in which the accelerating pace of plastic production loads the environment with ever more microplastics and nanoplastics year over year, a compounding and increasingly crushing debt. More plants and animals will die, ecosystems will buckle, and our bodies will grow ever lousier with plastic—the descent into ethical and environmental bankruptcy.

But another path is reminding ourselves that in the grand scheme of human existence, it wasn't that long ago that we got along just fine without plastic. And the future isn't such a neat binary: plastic or no plastic. Plenty of societies moved seamlessly between the hunter-gatherer life and planting little plots of crops, whichever suited the conditions and their whims. It wasn't just one day *boom, civilization*, but a gradual creep. We are now at a similar crossroads, where we can lock in a future wrapped in plastic or we can pump the brakes. There's a path in which we rein in single-use packaging, fix the busted economics of recycling, and get a microfiber filter in every washing machine. Polymers can keep on serving essential purposes in planes and cars and medicine and sure, because it's what Michael Phelan would have wanted, a game of billiards. All the pieces of plastic great and small are out there for good, but we can at least turn down the tap.

# Acknowledgments

A big thanks to my editor at Island Press, Emily Turner, first of all for taking a chance on a heavy topic and for your shaping of the book. And as always, gratitude to David Fugate, agent extraordinaire.

Much appreciation to Deonie and Steve Allen, who spent an inordinate amount of time on Zoom, even while aboard a boat, and to Janice Brahney for the blueberries and patience. A very big thanks (I've now run out of synonyms for thanks) to the hundred or so other researchers who interviewed for this book, many of whom were not mentioned in the text, so I want to acknowledge: Frank Pierik, Irina Rypina, Madelaine Bourdages, Garth Covernton, Elora Fournier, Lucie Etienne-Mesmin, Charlotte Grootaert, Kristian Syberg, Stéphanie Blanquet-Diot, Muriel Mercier-Bonin, Jonathan Millican, Seema Agarwal, Laura Markley, Maxine Swee-Li Yee, Chee-Onn Leong, Nick Beijer, Rudolf Reuther, Mariacristina Cocca, Helene Wiesinger, Lauren Divine, Peter Strohriegl, Britta Baechler, Michael LeVine, Gerhard Rambold, Stephan Rohrbach, Marcus Horn, Melanie Pöhlmann, Roman Lehner, Max Beaurepaire, Parisa Ariya, Dick Vethaak, Kurunthachalam Kannan, Mojca Zupan, Andrej Kržan, Jeroen Dagevos, Veronica Padula, Dušan

Materić, Holger Kress, Joana Prata, Nia Jones, Helen Ford, Emily Miller, Kelly D. Moran, Rebecca Sutton, Amanda Dawson, Bethanie Carney Almroth, and Patricia Villarrubia-Gómez.

Thank you, Mom and Dad and the Gambles too—I may not have kids of my own, but I may as well try to make the world a better place for you all. And Grandma, for the hand-me-down Roomba toiling to rid my house of microplastics.

# Notes


**Introduction**

1. Brahney, Janice, Margaret Hallerud, Eric Heim, Maura Hahnenberger, and Suja Sukumaran. 2020. "Plastic Rain in Protected Areas of the United States." *Science* 368:1257–60. https://doi.org/10.1126/science.aaz5819.
2. Simon, Matt. 2020. "Plastic Rain Is the New Acid Rain." *Wired*. https://www.wired.com/story/plastic-rain-is-the-new-acid-rain/.
3. Zheng, Jiajia, and Sangwon Suh. 2019. "Strategies to Reduce the Global Carbon Footprint of Plastics." *Nature Climate Change* 9:374–78. https://doi.org/10.1038/s41558-019-0459-z.

**Chapter 1: Welcome to Planet Plastic**

1. Robbett, Mary Kate. 2018. "Imitation Ivory and the Power of Play." Smithsonian National Museum of American History. https://invention.si.edu/imitation-ivory-and-power-play.
2. Altman, Rebecca. 2021. "The Myth of Historical Bio-Based Plastics." *Science* 373 (6550): 47–49. https://doi.org/10.1126/science.abj1003; Freinkel, Susan. 2011. *Plastic: A Toxic Love Story*. New York: Houghton Mifflin Harcourt.
3. Science Museum. 2019. "The Age of Plastic: From Parkesine to Pollution." https://www.sciencemuseum.org.uk/objects-and-stories/chemistry/age-plastic-parkesine-pollution.

4. American Chemical Society National Historic Chemical Landmarks. n.d. "Bakelite: The World's First Synthetic Plastic." http://www.acs.org/content/acs/en/education/whatischemistry/landmarks/bakelite.html.
5. Science History Institute. n.d. "History and Future of Plastics." https://www.sciencehistory.org/the-history-and-future-of-plastics.
6. Nicholson, Joseph L., and G. R. Leighton. 1942. "Plastics Come of Age." *Harper's Magazine*. August 1942.
7. *Life Magazine*. 1955. "Throwaway Living: Disposable Items Cut Down Household Chores." August 1, 1955.
8. OceanCare. 2016. "Microplastics Factsheet." https://oceancare.org/wp-content/uploads/2016/07/Factsheet_Mikroplastik_EN_2015.pdf.
9. Sundt, Peter, Per-Erik Schulze, and Frode Syversen. 2014. "Sources of Microplastic- Pollution to the Marine Environment." Mepex. https:// d3n8a8pro7vhmx.cloudfront.net/boomerangalliance/pages/507/attachments/original/1481155578/Norway_Sources_of_Microplastic_Pollution.pdf?1481155578.
10. Bashir, Saidu M., Sam Kimiko, Chu-Wa Mak, James Kar-Hei Fang, and David Gonçalves. 2021. "Personal Care and Cosmetic Products as a Potential Source of Environmental Contamination by Microplastics in a Densely Populated Asian City." *Frontiers in Marine Science* 8. https://doi.org/10.3389/fmars.2021.683482; Kramm, Johanna, and Carolin Volker. 2018. "Understanding the Risks of Microplastics: A Social-Ecological Risk Perspective." In *Freshwater Microplastics: Emerging Environmental Contaminants?* The Handbook of Environmental Chemistry 58, edited by Martin Wagner and Scott Lambert. https://doi.org/10.1007/978-3-319-61615-5_11; Kleida, Danae. 2019. "Get to Know Microplastics in Your Cosmetics." Beat the Microbead: Plastic Soup Foundation. https://www.beatthemicrobead.org/get/get-to-know-microplastics-in-your-cosmetics-2/.
11. Ayodeji, Amobonye, Bhagwat Prashant, Raveendran Sindhu, Singh Suren, and Pillai Santhosh. 2021. "Environmental Impacts of Microplastics and Nanoplastics: A Current Overview." *Frontiers in Microbiology* 12:3728. https://doi.org/10.3389/fmicb.2021.768297.
12. Sun, Qing, Shu-Yan Ren, and Hong-Gang Ni. 2020. "Incidence of Microplastics in Personal Care Products: An Appreciable Part of Plastic Pollution." *Science of the Total Environment* 742:140218. https://doi.org/10.1016/j.scitotenv.2020.140218.
13. Cheung, Pui Kwan, and Lincoln Fok. 2017. "Characterisation of Plastic Microbeads in Facial Scrubs and Their Estimated Emissions in

Mainland China." *Water Research* 122:53–61. https://doi.org/10.1016/j.watres.2017.05.053.

14. Geyer, Roland, Jenna R. Jambeck, and Kara Lavender Law. 2017. "Production, Use, and Fate of All Plastics Ever Made." *Science Advances* 3 (7): e1700782. https://doi.org/10.1126/sciadv.1700782; Simon, Matt. 2020. "All the Stuff Humans Make Now Outweighs Earth's Organisms." *Wired.* https://www.wired.com/story/all-the-stuff-humans-make-now-outweighs-earths-organisms/.
15. Gross, Michael. 2017. "Our Planet Wrapped in Plastic." *Current Biology* 27 (16). https://doi.org/10.1016/j.cub.2017.08.007.
16. World Economic Forum. 2016. "The New Plastics Economy: Rethinking the Future of Plastics." https://ellenmacarthurfoundation.org/the-new-plastics-economy-rethinking-the-future-of-plastics.
17. Boucher, Julien, and Damien Friot. 2017. *Primary Microplastics in the Oceans: A Global Evaluation of Sources.* Gland, Switzerland: International Union for Conservation of Nature.
18. UNEP (United Nations Environment Programme). 2018. "Single-Use Plastics: A Roadmap for Sustainability." https://www.unep.org/resources/report/single-use-plastics-roadmap-sustainability.
19. National Academies of Sciences, Engineering, and Medicine. 2021. *Reckoning with the U.S. Role in Global Ocean Plastic Waste.* Washington, DC: National Academies Press. https://doi.org/10.17226/26132.
20. Gardiner, Beth. 2019. "The Plastics Pipeline: A Surge of New Production Is on the Way." Yale Environment 360. https://e360.yale.edu/features/the-plastics-pipeline-a-surge-of-new-production-is-on-the-way.
21. UNEP. 2021. *From Pollution to Solution: A Global Assessment of Marine Litter and Plastic Pollution.* Nairobi: UNEP. https://www.unep.org/resources/pollution-solution-global-assessment-marine-litter-and-plastic-pollution; Lau, Winnie, Yonathan Shiran, Richard M. Bailey, Ed Cook, Martin R. Stuchtey, Julia Koskella, Costas A. Velis, et al. 2020. "Evaluating Scenarios Toward Zero Plastic Pollution." *Science* 369 (6510). https://doi.org/10.1126/science.aba9475.
22. Sharpe, Pete. 2015. "Making Plastics: From Monomer to Polymer." American Institute of Chemical Engineers. https://www.aiche.org/resources/publications/cep/2015/september/making-plastics-monomer-polymer.
23. Center for International Environmental Law. 2019. "Plastic and Climate: The Hidden Costs of a Plastic Planet." https://www.ciel.org/plasticandclimate/.

24. Beyond Plastics. 2021. "The New Coal: Plastics and Climate Change." https://www.beyondplastics.org/plastics-and-climate.
25. Royer, Sarah-Jeanne, Sara Ferrón, Samuel T. Wilson, and David M. Karl. 2018. "Production of Methane and Ethylene From Plastic in the Environment." *PLoS ONE* 13 (8): e0200574. https://doi.org/10.1371/journal.pone.0200574.
26. Narancic, Tanja, Steven Verstichel, Srinivasa Reddy Chaganti, Laura Morales-Gamez, Shane T. Kenny, Bruno De Wilde, Ramesh Babu Padamati, and Kevin E. O'Connor. 2018. "Biodegradable Plastic Blends Create New Possibilities for End-of-Life Management of Plastics but They Are Not a Panacea for Plastic Pollution." *Environmental Science and Technology* 52 (18): 10441–52. https://doi.org/10.1021/acs.est.8b02963.
27. Royal Society Te Apārangi. 2019. "Plastics in the Environment." https://www.royalsociety.org.nz/major-issues-and-projects/plastics.
28. Lambert, Scott, and Martin Wagner. 2017. "Environmental Performance of Bio-Based and Biodegradable Plastics: The Road Ahead." *Chemical Society Reviews* 46 (22): 6855–71. https://doi.org/10.1039/C7CS00149E.
29. Napper, Imogen E., and Richard C. Thompson. 2019. "Environmental Deterioration of Biodegradable, Oxo-Biodegradable, Compostable, and Conventional Plastic Carrier Bags in the Sea, Soil, and Open-Air over a 3-Year Period." *Environmental Science and Technology* 53 (9): 4775–83. https://doi.org/10.1021/acs.est.8b06984.
30. Liao, Jin, and Qiqing Chen. 2021. "Biodegradable Plastics in the Air and Soil Environment: Low Degradation Rate and High Microplastics Formation." *Journal of Hazardous Materials* 418:126329. https://doi.org/10.1016/j.jhazmat.2021.126329.
31. Zimmermann, Lisa, Andrea Dombrowski, Carolin Völker, and Martin Wagner. 2020. "Are Bioplastics and Plant-Based Materials Safer Than Conventional Plastics? In Vitro Toxicity and Chemical Composition." *Environment International* 145:106066. https://doi.org/10.1016/j.envint.2020.106066.
32. Rillig, Matthias C., Shin Woong Kim, Tae-Young Kim, and Walter R. Waldman. 2021. "The Global Plastic Toxicity Debt." *Environmental Science and Technology* 55, (5): 2717–19. https://doi.org/10.1021/acs.est.0c07781.
33. Halden, Rolf U., Charles Rolsky, and Farhan R. Khan. 2021. "Time: A Key Driver of Uncertainty When Assessing the Risk of Environmental

Plastics to Human Health." *Environmental Science and Technology* 55 (19): 12766–69. https://doi.org/10.1021/acs.est.1c02580.

34. Zimmermann, Lisa, Georg Dierkes, Thomas A. Ternes, Carolin Völker, and Martin Wagner. 2019. "Benchmarking the in Vitro Toxicity and Chemical Composition of Plastic Consumer Products." *Environmental Science and Technology* 53 (19): 11467–77. https://doi.org/10.1021/acs.est.9b02293; American Chemistry Council. n.d. "High Phthalates." https://www.americanchemistry.com/chemistry-in-america/chemistries/high-phthalates.
35. Stenmarck, Åsa, Elin L. Belleza, Anna Fråne, Niels Busch, Åge Larsen, and Margareta Wahlström. 2017. "Hazardous Substances in Plastics—Ways to Increase Recycling." Nordic Council of Ministers. http://norden.diva-portal.org/smash/record.jsf?pid=diva2%3A1070548&dswid=7787.
36. Turner, Andrew, and Montserrat Filella. 2021. "Hazardous Metal Additives in Plastics and Their Environmental Impacts." *Environment International* 156:106622. https://doi.org/10.1016/j.envint.2021.106622.
37. Muncke, Jane, Anna-Maria Andersson, Thomas Backhaus, Justin M. Boucher, Bethanie Carney Almroth, Arturo Castillo Castillo, Jonathan Chevrier, et al. 2020. "Impacts of Food Contact Chemicals on Human Health: a Consensus Statement." *Environmental Health* 19. https://doi.org/10.1186/s12940-020-0572-5; Azoulay, David, Priscilla Villa, Yvette Arellano, Miriam Gordon, Doun Moon, Kathryn Miller, and Kristen Thompson. 2019. "Plastic and Health: The Hidden Costs of a Plastic Planet." https://www.ciel.org/reports/plastic-health-the-hidden-costs-of-a-plastic-planet-february-2019/.
38. Walsh, Anna N., Christopher M. Reddy, Sydney F. Niles, Amy M. McKenna, Colleen M. Hansel, and Collin P. Ward. 2021. "Plastic Formulation Is an Emerging Control of Its Photochemical Fate in the Ocean." *Environmental Science and Technology* 55 (18): 12383–92. https://doi.org/10.1021/acs.est.1c02272; Woods Hole Oceanographic Institution. 2021. "Sunlight Can Break Down Marine Plastic into Tens of Thousands of Chemical Compounds, Study Finds." https://www.whoi.edu/press-room/news-release/sunlight-can-break-down-marine-plastic-into-tens-of-thousands-of-chemical-compounds-study-finds/.
39. Ward, Collin P., and Christopher M. Reddy. 2020. "Opinion: We Need Better Data about the Environmental Persistence of Plastic Goods." *Proceedings of the National Academy of Sciences*. 117 (26): 14618–21. https://doi.org/10.1073/pnas.2008009117.

40. Meides, Nora, Teresa Menzel, Björn Poetzschner, Martin G. J. Löder, Ulrich Mansfeld, Peter Strohriegl, Volker Altstaedt, and Jürgen Senker. 2021. "Reconstructing the Environmental Degradation of Polystyrene by Accelerated Weathering." *Environmental Science and Technology* 55 (12): 7930–38. https://doi.org/10.1021/acs.est.0c07718.
41. Hartmann, Nanna B., Thorsten Hüffer, Richard C. Thompson, Martin Hassellöv, Anja Verschoor, Anders E. Daugaard, Sinja Rist, et al. 2019. "Are We Speaking the Same Language? Recommendations for a Definition and Categorization Framework for Plastic Debris." *Environmental Science and Technology* 53 (3): 1039–47. https://doi.org/10.1021/acs.est.8b05297.
42. Arthur, Courtney, Joel Baker, and Holly Bamford. 2009. "Proceedings of the International Research Workshop on the Occurrence, Effects and Fate of Microplastic Marine Debris." NOAA Technical Memorandum NOS-OR&R-30. https://marinedebris.noaa.gov/proceedings-international-research-workshop-microplastic-marine-debris.
43. Ter Halle, Alexandra, Laurent Jeanneau, Marion Martignac, Emilie Jardé, Boris Pedrono, Laurent Brach, and Julien Gigault. 2017. "Nanoplastic in the North Atlantic Subtropical Gyre." *Environmental Science and Technology* 51 (23): 13689–97. https://doi.org/10.1021/acs.est.7b03667; Wayman, Chloe, and Helge Niemann. 2021. "The Fate of Plastic in the Ocean Environment: A Minireview." *Environmental Science Processes and Impacts* 23:198. https://doi.org/10.1039/D0EM00446D.
44. Mitrano, Denise, Peter Wick, and Bernd Nowack. 2021. "Placing Nanoplastics in the Context of Global Plastic Pollution." *Nature Nanotechnology* 16:491–500. https://doi.org/10.1038/s41565-021-00888-2; Gigault, Julien, Hind El Hadri, Brian Nguyen, Bruno Grassl, Laura Rowenczyk, Nathalie Tufenkji, Siyuan Feng, and Mark Wiesner. 2021. "Nanoplastics Are Neither Microplastics nor Engineered Nanoparticles." *Nature Nanotechnology* 16:501–7. https://doi.org/10.1038/s41565-021-00886-4.
45. Simon-Sánchez, Laura, Michaël Grelaud, Marco Franci, and Patrizia Ziveri. 2022. "Are Research Methods Shaping Our Understanding of Microplastic Pollution? A Literature Review on the Seawater and Sediment Bodies of the Mediterranean Sea." *Environmental Pollution* 292 (B): 118275. https://doi.org/10.1016/j.envpol.2021.118275.
46. Picó, Yolanda, Rodrigo Alvarez-Ruiz, Ahmed H. Alfarhan, Mohamed A. El-Sheikh, Hamad O. Alshahrani, and Damià Barceló. 2020.

"Pharmaceuticals, Pesticides, Personal Care Products and Microplastics Contamination Assessment of Al-Hassa Irrigation Network (Saudi Arabia) and Its Shallow Lakes." *Science of the Total Environment* 701. https://doi.org/10.1016/j.scitotenv.2019.135021.
47. Schlanger, Zoë. 2019. "Virgin Plastic Pellets Are the Biggest Pollution Disaster You've Never Heard Of." Quartz. https://qz.com/1689529/nurdles-are-the-biggest-pollution-disaster-youve-never-heard-of/.
48. Johnson, Chloe. 2021. "SC Plastic Pellet Spill Lawsuit Settled for $1 Million." *Post and Courier* (Charleston). https://www.postandcourier.com/news/sc-plastic-pellet-spill-lawsuit-settled-for-1-million/article_5330d994-7b96-11eb-b4e7-b7b8ca40fc27.html.
49. Fernández, Stacy. 2019. "Plastic Company Set to Pay $50 Million Settlement in Water Pollution Suit Brought on by Texas Residents." *Texas Tribune* (Austin). https://www.texastribune.org/2019/10/15/formosa-plastics-pay-50-million-texas-clean-water-act-lawsuit/.
50. NOAA Office for Coastal Management. n.d. "Historic Pollution Settlement Awards $1 Million to Nurdle Patrol." https://coast.noaa.gov/states/stories/historic-pollution-settlement-to-nurdle-patrol.html.
51. Reuters. 2012. "Hong Kong Government Criticized over Plastic Spill on Beaches." https://www.reuters.com/article/us-hongkong-spill/hong-kong-government-criticized-over-plastic-spill-on-beaches-idUSBRE87306J20120805.
52. Partow, Hassan, Camille Lacroix, Stephane Le Floch, and Luigi Alcaro. 2021. "X-Press Pearl Maritime Disaster: Sri Lanka—Report of the UN Environmental Advisory Mission." UNEP. https://wedocs.unep.org/handle/20.500.11822/36608?show=full; McVeigh, Karen. 2021. "Nurdles: The Worst Toxic Waste You've Probably Never Heard Of." *Guardian*. https://www.theguardian.com/environment/2021/nov/29/nurdles-plastic-pellets-environmental-ocean-spills-toxic-waste-not-classified-hazardous.
53. De Vos, Asha, Lihini Aluwihare, Sarah Youngs, Michelle H. DiBenedetto, Collin P. Ward, Anna P. M. Michel, Beckett C. Colson, et al. 2021. "The M/V X-Press Pearl Nurdle Spill: Contamination of Burnt Plastic and Unburnt Nurdles along Sri Lanka's Beaches." *ACS Environmental Au*. https://doi.org/10.1021/acsenvironau.1c00031.
54. Erni-Cassola, Gabriel, Vinko Zadjelovic, Matthew I. Gibson, and Joseph A. Christie-Oleza. 2019. "Distribution of Plastic Polymer Types in the Marine Environment: A Meta-Analysis." *Journal of Hazardous Materials* 369:691–98. https://doi.org/10.1016/j.jhazmat.2019.02.067.

55. The Great Nurdle Hunt. n.d. "The Problem." https://www.nurdlehunt.org.uk/the-problem.html.
56. Lozano, Rebeca Lopez, and John Mouat. 2009. "Marine Litter in the North-East Atlantic Region: Assessment and Priorities for Response." KIMO International.
57. Turra, Alexander, Aruanã B. Manzano, Rodolfo Jasão S. Dias, Michel M. Mahiques, Lucas Barbosa, Danilo Balthazar-Silva, and Fabiana T. Moreira. 2014. "Three-Dimensional Distribution of Plastic Pellets in Sandy Beaches: Shifting Paradigms." *Scientific Reports* 4:4435. https://doi.org/10.1038/srep04435.
58. Abu-Hilal, Ahmad H., and Tariq H. Al-Najjar. 2009. "Plastic Pellets on the Beaches of the Northern Gulf of Aqaba, Red Sea." *Aquatic Ecosystem Health and Management* 12 (4): 461–70. https://doi.org/10.1080/14634980903361200.
59. Karlsson, Therese M., Lars Arneborg, Göran Broström, Bethanie Carney Almroth, Lena Gipperth, and Martin Hassellöv. 2018. "The Unaccountability Case of Plastic Pellet Pollution." *Marine Pollution Bulletin* 129 (1): 52–60. https://doi.org/10.1016/j.marpolbul.2018.01.041; Fidra. n.d. *Study to Quantify Plastic Pellet Loss in the UK: Report Briefing*. East Lothian, Scotland. http://www.nurdlehunt.org.uk/images/Leaflets/Report_briefing.pdf.
60. Cole, Matthew, Pennie Lindeque, Claudia Halsband, and Tamara S. Galloway. 2011. "Microplastics as Contaminants in the Marine Environment: A Review." *Marine Pollution Bulletin* 62 (12): 2588–97. https://doi.org/10.1016/j.marpolbul.2011.09.025.
61. Rochman, Chelsea, Cole Brookson, Jacqueline Bikker, Natasha Djuric, Arielle Earn, Kennedy Bucci, Samantha Athey, et al. 2019. "Rethinking Microplastics as a Diverse Contaminant Suite." *Environmental Toxicology and Chemistry* 38 (4): 703–11. https://doi.org/10.1002/etc.4371.
62. Environmental Protection Agency (EPA). 1998. "How Wastewater Treatment Works." https://www3.epa.gov/npdes/pubs/bastre.pdf.
63. Altreuter, Róisín Magee. 2017. *Microfibers, Macro Problems*. Los Angeles: 5 Gyres Institute. https://static1.squarespace.com/static/5522e85be4b0b65a7c78ac96/t/5a66456cc83025f49135fbc2/1516651904354/Microfibers%2C+Macro+problems.pdf.
64. Ellen MacArthur Foundation. 2017. "A New Textiles Economy: Redesigning Fashion's Future." https://ellenmacarthurfoundation.org/a-new-textiles-economy.

65. UNEP. 2019. "Fashion's Tiny Hidden Secret." https://www.unep.org/news-and-stories/story/fashions-tiny-hidden-secret.
66. Athey, Samantha, and Lisa Erdle. 2021. "Are We Underestimating Anthropogenic Microfiber Pollution? A Critical Review of Occurrence, Methods and Reporting." *Environmental Toxicology and Chemistry*. https://doi.org/10.1002/etc.5173.
67. McIlwraith, Hayley K., Jack Lin, Lisa M. Erdle, Nicholas Mallos, Miriam L. Diamond, and Chelsea M. Rochman. 2019. "Capturing Microfibers: Marketed Technologies Reduce Microfiber Emissions from Washing Machines." *Marine Pollution Bulletin* 139:40–45. https://doi.org/10.1016/j.marpolbul.2018.12.012.
68. Napper, Imogen E., and Richard C. Thompson. 2016. "Release of Synthetic Microplastic Plastic Fibres from Domestic Washing Machines: Effects of Fabric Type and Washing Conditions." *Marine Pollution Bulletin* 112 (1–2): 39–45. https://doi.org/10.1016/j.marpolbul.2016.09.025.
69. Almroth, Bethanie M. Carney, Linn Åström, Sofia Roslund, Hanna Petersson, Mats Johansson, and Nils-Krister Persson. 2018. "Quantifying Shedding of Synthetic Fibers From Textiles: A Source of Microplastics Released into the Environment." *Environmental Science and Pollution Research* 25:1191–99. https://doi.org/10.1007/s11356-017-0528-7.
70. De Falco, Francesca, Maria Pia Gulloa, Gennaro Gentile, Emilia Di Pace, Mariacristina Cocca, Laura Gelabert, Marolda Brouta-Agnésa, et al. 2018. "Evaluation of Microplastic Release Caused by Textile Washing Processes of Synthetic Fabrics." *Environmental Pollution* 236:916–25. https://doi.org/10.1016/j.envpol.2017.10.057.
71. Yang, Tong, Jialuo Luo, and Bernd Nowack. 2021. "Characterization of Nanoplastics, Fibrils, and Microplastics Released during Washing and Abrasion of Polyester Textiles." *Environmental Science and Technology* 55 (23): 15873–81. https://doi.org/10.1021/acs.est.1c04826.
72. Murphy, Fionn, Ciaran Ewins, Frederic Carbonnier, and Brian Quinn. 2016. "Wastewater Treatment Works (WwTW) as a Source of Microplastics in the Aquatic Environment." *Environmental Science and Technology* 50 (11): 5800–5808. https://doi.org/10.1021/acs.est.5b05416.
73. Conley, Kenda, Allan Clum, Jestine Deepe, Haven Lane, and Barbara Beckingham. 2019. "Wastewater Treatment Plants as a Source of Microplastics to an Urban Estuary: Removal Efficiencies and Loading per Capita over One Year." *Water Research X* 3:100030. https://doi.org/10.1016/j.wroa.2019.100030.

74. Utrecht University. 2021. "Half of Global Wastewater Treated, Rates in Developing Countries Still Lagging." *Science Daily*. https://www.sciencedaily.com/releases/2021/02/210208085457.htm.
75. Jones, Edward R., Michelle T. H. van Vliet, Manzoor Qadir, and Marc F. P. Bierkens. 2021. "Country-Level and Gridded Estimates of Wastewater Production, Collection, Treatment and Reuse." *Earth System Science Data* 13 (2): 237. https://doi.org/10.5194/essd-13-237-2021.
76. Gavigan, Jenna, Timnit Kefela, Ilan Macadam-Somer, Sangwon Suh, and Roland Geyer. 2020. "Synthetic Microfiber Emissions to Land Rival Those to Waterbodies and Are Growing." *PLoS ONE* 15 (9): e0237839. https://doi.org/10.1371/journal.pone.0237839.
77. European Parliament. 2020."Plastic Microfibre Filters for New Washing Machines by 2025." https://www.europarl.europa.eu/doceo/document/E-9-2020-001371_EN.html.
78. Geyer, Roland, Jenna Gavigan, Alexis M. Jackson, Vienna R. Saccomanno, Sangwon Suh, and Mary G. Gleason. 2022. "Quantity and Fate of Synthetic Microfiber Emissions from Apparel Washing in California and Strategies for Their Reduction." *Environmental Pollution* 298:118835. https://doi.org/10.1016/j.envpol.2022.118835.
79. Dempsey, Tom. 2021. "Toward Eliminating Pre-Consumer Emissions of Microplastics from the Textile Industry." Nature Conservancy and Bain and Company. https://www.nature.org/content/dam/tnc/nature/en/documents/210322TNCBain_Pre-ConsumerMicrofiberEmissionsv6.pdf; Simon, Matt. 2021. "Your Clothes Spew Microfibers before They're Even Clothes." *Wired*. https://www.wired.com/story/your-clothes-spew-microfibers-before-theyre-even-clothes/.
80. Microfibre Consortium. n.d. "Our Signatories to The Microfibre 2030 Commitment." https://www.microfibreconsortium.com/signatories.
81. Vassilenko, Ekaterina, Mathew Watkins, Stephen Chastain, Joel Mertens, Anna M. Posacka, Shreyas Patankar, and Peter S. Ross. 2021. "Domestic Laundry and Microfiber Pollution: Exploring Fiber Shedding from Consumer Apparel Textiles." *PLoS ONE* 16 (7): e0250346. https://doi.org/10.1371/journal.pone.0250346.
82. Khan, Anum, Barry Orr, and Darko Joksimovic. 2019. "Defining 'Flushability' for Sewer Use: Ryerson Urban Water." https://www.ryerson.ca/water/research/flushability/.
83. Ó Briain, Oisín, Ana R. Marques Mendes, Stephen McCarron, Mark G. Healy, and Liam Morrison. 2020. "The Role of Wet Wipes and Sanitary Towels as a Source of White Microplastic Fibres in the Marine

Environment." *Water Research* 182:116021. https://doi.org/10.1016/j.watres.2020.116021.

**Chapter 2: A Voyage on the Synthetic Seas**

1. Carpenter, Edward J., and K. L. Smith. 1972. "Plastics on the Sargasso Sea Surface." *Science* 175 (4027): 1240–41. https://doi.org/10.1126/science.175.4027.1240; Ryan, Peter. 2015. "A Brief History of Marine Litter Research." In *Marine Anthropogenic Litter*, edited by Melanie Bergmann, Lars Gutow, and Michael Klages. N.p.: Springer. https://doi.org/10.1007/978-3-319-16510-3_1.
2. Frias, J. P. G. L., and Roisin Nash. 2019. "Microplastics: Finding a Consensus on the Definition." *Marine Pollution Bulletin* 138:145–47. https://doi.org/10.1016/j.marpolbul.2018.11.022.
3. Carpenter, Edward J., Susan J. Anderson, George R. Harvey, Helen P. Miklas, and Bradford B. Peck. 1972. "Polystyrene Spherules in Coastal Waters." *Science* 178 (4062): 749–50. https://doi.org/10.1126/science.178.4062.749.
4. Rensberger, Boyce. 1972. "Plastic Is Found in the Sargasso Sea." *New York Times,* March 19, 1972.
5. Eriksen, Marcus, Laurent C. M. Lebreton, Henry S. Carson, Martin Thiel, Charles J. Moore, Jose C. Borerro, Francois Galgani, Peter G. Ryan, and Julia Reisser. 2014. "Plastic Pollution in the World's Oceans: More than 5 Trillion Plastic Pieces Weighing Over 250,000 Tons Afloat at Sea." *PLoS ONE* 9 (12): e111913. https://doi.org/10.1371/journal.pone.0111913.
6. Pabortsava, Katsiaryna, and Richard S. Lampitt. 2020. "High Concentrations of Plastic Hidden Beneath the Surface of the Atlantic Ocean." *Nature Communications* 11:4073. https://doi.org/10.1038/s41467-020-17932-9; Simon, Matt. 2020. "Wait, How Much Microplastic Is Swirling in the Atlantic?" *Wired.* https://www.wired.com/story/how-much-microplastic-is-swirling-in-the-atlantic/.
7. Simon, Matt. 2019. "Monterey Bay Is a Natural Wonder—Poisoned with Microplastic." *Wired.* https://www.wired.com/story/monterey-bay-microplastic/; Choy, C. Anela, Bruce H. Robison, Tyler O. Gagne, Benjamin Erwin, Evan Firl, Rolf U. Halden, J. Andrew Hamilton, et al. "The Vertical Distribution and Biological Transport of Marine Microplastics across the Epipelagic and Mesopelagic Water Column." *Scientific Reports* 9:7843. https://doi.org/10.1038/s41598-019-44117-2.
8. Brandon, Jennifer A., Alexandra Freibott, and Linsey M. Sala. 2020.

"Patterns of Suspended and Salp-Ingested Microplastic Debris in the North Pacific Investigated with Epifluorescence Microscopy." *Limnology and Oceanography Letters* 5 (1). https://doi.org/10.1002/lol2.10127.

9. Isobe, Atsuhiko, Takafumi Azuma, Muhammad Reza Cordova, Andrés Cózar, Francois Galgani, Ryuichi Hagita, La Daana Kanhai, et al. 2021. "A Multilevel Dataset of Microplastic Abundance in the World's Upper Ocean and the Laurentian Great Lakes." *Microplastics and Nanoplastics* 1:16. https://doi.org/10.1186/s43591-021-00013-z.
10. Franklin Rey, Savannah, Janet Franklin, and Sergio J. Rey. 2021. "Microplastic Pollution on Island Beaches, Oahu, Hawaii." *PLoS ONE* 16 (2): e0247224. https://doi.org/10.1371/journal.pone.0247224.
11. Hidalgo-Ruz, Valeria, and Martin Thiel. 2013. "Distribution and Abundance of Small Plastic Debris on Beaches in the SE Pacific (Chile): A Study Supported by a Citizen Science Project." *Marine Environmental Research* 87–88:12–18. https://doi.org/10.1016/j.marenvres.2013.02.015; Jones, Jen S., Adam Porter, Juan Pablo Muñoz-Pérez, Daniela Alarcón-Ruales, Tamara S. Galloway, Brendan J. Godleye, David Santillo, et al. 2021. "Plastic Contamination of a Galapagos Island (Ecuador) and the Relative Risks to Native Marine Species." *Science of the Total Environment* 789:147704. https://doi.org/10.1016/j.scitotenv.2021.147704.
12. Lavers, Jennifer L., and Alexander L. Bond. 2017. "Exceptional and Rapid Accumulation of Anthropogenic Debris on One of the World's Most Remote and Pristine Islands." *Proceedings of the National Academy of Sciences* 114 (23): 6052–55. https://doi.org/10.1073/pnas.1619818114.
13. Lavers, Jennifer L., Jack Rivers-Auty, and Alexander L. Bond. 2021. "Plastic Debris Increases Circadian Temperature Extremes in Beach Sediments." *Journal of Hazardous Materials* 416:126140. https://doi.org/10.1016/j.jhazmat.2021.126140.
14. Guarino, Ben. 2018. "Climate Change Is Turning 99 Percent of These Baby Sea Turtles Female." *Washington Post*. https://www.washingtonpost.com/news/speaking-of-science/wp/2018/01/08/climate-change-is-turning-99-percent-of-these-baby-sea-turtles-female/.
15. Zhang, Ting, Liu Lin, Deqin Li, Shannan Wu, Li Kong, Jichao Wang, and Haitao Shi. 2021. "The Microplastic Pollution in Beaches That Served as Historical Nesting Grounds for Green Turtles on Hainan Island, China." *Marine Pollution Bulletin* 173 (B): 113069. https://doi.org/10.1016/j.marpolbul.2021.113069.

16. Kane, Ian A., Michael A. Clare, Elda Miramontes, Roy Wogelius, James J. Rothwell, Pierre Garreau, and Florian Pohl. 2020. "Seafloor Microplastic Hotspots Controlled by Deep-Sea Circulation." *Science* 368 (6495): 1140–45. https://doi.org/10.1126/science.aba5899.
17. Simon, Matt. 2020. "'Microplastic Hot Spots' Are Tainting Deep-Sea Ecosystems." *Wired.* https://www.wired.com/story/microplastic-hotspots/.
18. European Environmental Agency. 2015. "Mediterranean Sea Region Briefing: The European Environment—State and Outlook 2015." https://www.eea.europa.eu/soer/2015/countries/mediterranean.
19. United Nations Environment Programme. 2010. *Assessment of the State of Microbial Pollution in the Mediterranean Sea.* Mediterranean Action Plan Technical Reports, Series No. 170. Athens: UNEP/MAP. https://wedocs.unep.org/bitstream/handle/20.500.11822/520/mts170.pdf?sequence=2.
20. Abreu, André, and Maria Luiza Pedrotti. 2019. "Microplastics in the Oceans: The Solutions Lie on Land." *Field Actions Science Reports* (special issue 19): 62–67. http://journals.openedition.org/factsreports/5290.
21. Kuczenski, Brandon, Camila Vargas Poulsen, Eric L. Gilman, Michael Musyl, Roland Geyer, and Jono Wilson. 2021. "Plastic Gear Loss Estimates from Remote Observation of Industrial Fishing Activity." *Fish and Fisheries* 23 (1): 22–33. https://doi.org/10.1111/faf.12596.
22. Dowarah, Kaushik, and Suja P. Devipriya. 2019. "Microplastic Prevalence in the Beaches of Puducherry, India and Its Correlation with Fishing and Tourism/Recreational Activities." *Marine Pollution Bulletin* 148:123–33. https://doi.org/10.1016/j.marpolbul.2019.07.066.
23. Lusher, Amy, Peter Hollman, and Jeremy Mendoza-Hill. 2017. *Microplastics in Fisheries and Aquaculture: Status of Knowledge on Their Occurrence and Implications for Aquatic Organisms and Food Safety.* FAO Fisheries and Aquaculture Technical Paper No. 615. Rome: FAO. https://www.fao.org/documents/card/en/c/59bfa1fc-0875-4216-bd33-55b6003cfad8/.
24. Napper, Imogen Ellen, Luka Seamus Wright, Aaron C.Barrett, Florence N. F. Parker-Jurd, and Richard C. Thompson. 2022. "Potential Microplastic Release from the Maritime Industry: Abrasion of Rope." *Science of The Total Environment* 804:150155. https://doi.org/10.1016/j.scitotenv.2021.150155.
25. Dibke, Christopher, Marten Fischer, and Barbara M. Scholz-Böttcher. 2021. "Microplastic Mass Concentrations and Distribution in German

Bight Waters by Pyrolysis—Gas Chromatography—Mass Spectrometry/Thermochemolysis Reveal Potential Impact of Marine Coatings: Do Ships Leave Skid Marks?" *Environmental Science and Technology* 55 (4): 2285–95. https://doi.org/10.1021/acs.est.0c04522.

26. Gaylarde, Christine C., José Antonio Baptista Neto, and Estefan Monteiro da Fonseca. 2021. "Paint Fragments as Polluting Microplastics: A Brief Review." *Marine Pollution Bulletin* 162:111847. https://doi.org/10.1016/j.marpolbul.2020.111847; Cardozo, Ana L. P., Eduardo G. G. Farias, Jorge L. Rodrigues-Filho, Isabel B. Moteiro, Tatianny M. Scandolo, and David V. Dantas. 2018. "Feeding Ecology and Ingestion of Plastic Fragments by Priacanthus arenatus: What's the Fisheries Contribution to the Problem?" *Marine Pollution Bulletin* 130:19–27. https://doi.org/10.1016/j.marpolbul.2018.03.010.
27. Costamare Inc. n.d. "Container Facts." https://www.costamare.com/industry_containerisation.
28. Research and Markets. 2020. "Global Sea Freight Forwarding Market (2020 to 2025)—Growth, Trends, and Forecasts." https://www.globenewswire.com/news-release/2020/12/03/2138926/0/en/Global-Sea-Freight-Forwarding-Market-2020-to-2025-Growth-Trends-and-Forecasts.html.
29. Leistenschneider, Clara, Patricia Burkhardt-Holm, Thomas Mani, Sebastian Primpke, Heidi Taubner, and Gunnar Gerdts. 2021. "Microplastics in the Weddell Sea (Antarctica): A Forensic Approach for Discrimination between Environmental and Vessel-Induced Microplastics." *Environmental Science and Technology* 55 (23): 15900–11. https://doi.org/10.1021/acs.est.1c05207.
30. Peng, Guyu, Baile Xu, and Daoji Li. 2021. "Gray Water from Ships: A Significant Sea-Based Source of Microplastics?" *Environmental Science and Technology* 56 (1): 4–7. https://doi.org/10.1021/acs.est.1c05446.
31. Turner, Andrew. 2021. "Paint Particles in the Marine Environment: An Overlooked Component of Microplastics." *Water Research X* 12:100110. https://doi.org/10.1016/j.wroa.2021.100110.
32. Brandon, Jennifer A., William Jones, and Mark D. Ohman. 2019. "Multidecadal Increase in Plastic Particles in Coastal Ocean Sediments." *Science Advances* 5 (9): eaax0587. https://doi.org/10.1126/sciadv.aax0587; Simon, Matt. 2019. "Plastic Will Be the Shameful Artifact Our Descendants Dig Up." *Wired*. https://www.wired.com/story/microplastic-core-samples/.
33. Matsuguma, Yukari, Hideshige Takada, Hidetoshi Kumata, Hirohide Kanke, Shigeaki Sakurai, Tokuma Suzuki, Maki Itoh, et al. 2017.

"Microplastics in Sediment Cores from Asia and Africa as Indicators of Temporal Trends in Plastic Pollution." *Archives of Environmental Contamination and Toxicology* 73 (2): 230–39. https://doi.org/10.1007/s00244-017-0414-9; Martin, C., F. Baalkhuyur, L. Valluzzi, V. Saderne, M. Cusack, H. Almahasheer, P. K. Krishnakumar, et al. 2020. "Exponential Increase of Plastic Burial in Mangrove Sediments as a Major Plastic Sink." *Science Advances* 6 (44). https://doi.org/10.1126/sciadv.aaz5593.

34. Barrett, Justine, Zanna Chase, Jing Zhang, Mark M. Banaszak Holl, Kathryn Willis, Alan Williams, Britta D. Hardesty, and Chris Wilcox. 2020. "Microplastic Pollution in Deep-Sea Sediments from the Great Australian Bight." *Frontiers in Marine Science* 7. https://doi.org/10.3389/fmars.2020.576170.
35. Zalasiewicz, Jan, Colin N. Waters, Juliana A. Ivar do Sul, Patricia L. Corcoran, Anthony D. Barnosky, Alejandro Cearreta, Matt Edgeworth, et al. 2016. "The Geological Cycle of Plastics and Their Use as a Stratigraphic Indicator of the Anthropocene." *Anthropocene* 13:4–17. https://doi.org/10.1016/j.ancene.2016.01.002.
36. Ross, Peter, Stephen Chastain, Ekaterina Vassilenko, Anahita Etemadifar, Sarah Zimmermann, Sarah-Ann Quesnel, Jane Eert, et al. 2021. "Pervasive Distribution of Polyester Fibres in the Arctic Ocean Is Driven by Atlantic Inputs." *Nature Communications* 12:106. https://doi.org/10.1038/s41467-020-20347-1.
37. Simon, Matt. 2021. "The Arctic Ocean Is Teeming with Microfibers from Clothes." *Wired*. https://www.wired.com/story/the-arctic-ocean-is-teeming-with-microfibers-from-clothes/.
38. Athey, Samantha N., Jennifer K. Adams, Lisa M. Erdle, Liisa M. Jantunen, Paul A. Helm, Sarah A. Finkelstein, and Miriam L. Diamond. 2020. "The Widespread Environmental Footprint of Indigo Denim Microfibers from Blue Jeans." *Environmental Science and Technology Letters* 7 (11): 840–847. https://doi.org/10.1021/acs.estlett.0c00498.
39. Simon, Matt. 2020. "Your Beloved Blue Jeans Are Polluting the Ocean—Big Time." *Wired*. https://www.wired.com/story/your-blue-jeans-are-polluting-the-ocean/.
40. Arvai, Antonette. 2013. "More on IJC's Great Lakes Wastewater Treatment Study and Removing Chemicals of Emerging Concern." International Joint Commission. https://www.ijc.org/en/more-ijcs-great-lakes-wastewater-treatment-study-and-removing-chemicals-emerging-concern.
41. Voosen, Paul. 2021. "The Arctic Is Warming Four Times Faster Than

the Rest of the World." *Science News*. https://www.science.org/content/article/arctic-warming-four-times-faster-rest-world.

42. Obbard, Rachel W., Saeed Sadri, Ying Qi Wong, Alexandra A. Khitun, Ian Baker, and Richard C. Thompson. 2014. "Global Warming Releases Microplastic Legacy Frozen in Arctic Sea Ice." *Earth's Future* 2:315–20. https://doi.org/10.1002/2014EF000240.
43. Peeken, Ilka, Sebastian Primpke, Birte Beyer, Julia Gütermann, Christian Katlein, Thomas Krumpen, Melanie Bergmann, Laura Hehemann, and Gunnar Gerdts. 2018. "Arctic Sea Ice Is an Important Temporal Sink and Means of Transport for Microplastic." *Nature Communications* 9:1505. https://doi.org/10.1038/s41467-018-03825-5.
44. Katija, Kakani, C. Anela Choy, Rob E. Sherlock, Alana D. Sherman, Bruce H. Robison. 2017. "From the Surface to the Seafloor: How Giant Larvaceans Transport Microplastics into the Deep Sea." *Science Advances* 3 (8): e1700715. https://doi.org/10.1126/sciadv.1700715.
45. Niiler, Eric. 2017. "Plankton 'Mucus Houses' Could Pull Microplastics from the Sea." *Wired*. https://www.wired.com/story/plankton-mucus-houses-could-pull-microplastics-from-the-sea/.
46. Santos, Robson G., Gabriel E. Machovsky-Capuska, and Ryan Andrades. 2021. "Plastic Ingestion as an Evolutionary Trap: Toward a Holistic Understanding." *Science* 373 (6550): 56–60. https://doi.org/10.1126/science.abh0945.
47. Davison, Peter, and Rebecca G. Asch. 2011. "Plastic Ingestion by Mesopelagic Fishes in the North Pacific Subtropical Gyre." *Marine Ecology Progress Series* 432:173–80. https://doi.org/10.3354/meps09142.
48. Jamieson, Alan J., Lauren Brooks, William D. K. Reid, Stuart B. Piertney, and Bhavani E. Narayanaswamy. 2019. "Microplastics and Synthetic Particles Ingested by Deep-Sea Amphipods in Six of the Deepest Marine Ecosystems on Earth." *Royal Society Open Science* 6 (2): 1–11. https://doi.org/10.1098/rsos.180667.
49. Peng, Xiaotong, M. Chen, Shun Chen, Shamik Dasgupta, Hengchao Xu, Kaiwen Ta, Mengran Du, et al. 2018. "Microplastics Contaminate the Deepest Part of the World's Ocean." *Geochemical Perspectives Letters* 9:1–5. https://doi.org/10.7185/geochemlet.1829.
50. McGoran, Alexandra R., James S. Maclaine, Paul F. Clark, and David Morritt. 2021. "Synthetic and Semi-Synthetic Microplastic Ingestion by Mesopelagic Fishes from Tristan da Cunha and St Helena, South Atlantic." *Frontiers in Marine Science* 8:78. https://doi.org/10.3389/fmars.2021.633478.

51. Miller, Michaela E., Mark Hamann, and Frederieke J. Kroon. 2020."Bioaccumulation and Biomagnification of Microplastics in Marine Organisms: A Review and Meta-Analysis of Current Data." *PLoS ONE* 15 (10): e0240792. https://doi.org/10.1371/journal.pone.0240792.
52. University of Leicester. 2015. "Failing Phytoplankton, Failing Oxygen: Global Warming Disaster Could Suffocate Life on Planet Earth." *ScienceDaily*. https://www.sciencedaily.com/releases/2015/12/151201094120.htm.
53. Thompson, Richard C., Ylva Olsen, Richard P. Mitchell, Anthony Davis, Steven J. Rowland, Anthony W. G. John, Daniel McGonigle, and Andrea E. Russell. 2004. "Lost at Sea: Where Is All the Plastic?" *Science* 304 (5672): 838. https://doi.org/10.1126/science.1094559.
54. Absher, Theresinha Monteiro, Silvio Luiz Ferreira, Yargos Kern, Augusto Luiz Ferreira Jr., Susete Wambier Christo, and Rômulo Augusto Ando. 2019. "Incidence and Identification of Microfibers in Ocean Waters in Admiralty Bay, Antarctica." *Environmental Science and Pollution Research* 26:292–98. https://doi.org/10.1007/s11356-018-3509-6.
55. Lin, Vivian S. 2016. "Research Highlights: Impacts of Microplastics on Plankton." *Environmental Science: Processes and Impacts* 18:160–63. https://doi.org/10.1039/C6EM90004F; Desforges, Jean-Pierre W., Moira Galbraith, and Peter Ross. "Ingestion of Microplastics by Zooplankton in the Northeast Pacific Ocean." *Archives of Environmental Contamination and Toxicology* 69:320–30. https://doi.org/10.1007/s00244-015-0172-5.
56. Sun, Xiaoxia, Qingjie Li, Mingliang Zhu, Junhua Liang, Shan Zheng, and Yongfang Zhao. 2017. "Ingestion of Microplastics by Natural Zooplankton Groups in the Northern South China Sea." *Marine Pollution Bulletin* 115 (1–2): 217–24. https://doi.org/10.1016/j.marpolbul.2016.12.004.
57. Markic, Ana, Clarisse Niemand, James H. Bridson, Nabila Mazouni-Gaertner, Jean-Claude Gaertner, Marcus Eriksen, and Melissa Bowen. 2018. "Double Trouble in the South Pacific Subtropical Gyre: Increased Plastic Ingestion by Fish in the Oceanic Accumulation Zone." *Marine Pollution Bulletin* 136:547–64. https://doi.org/10.1016/j.marpolbul.2018.09.031.
58. Botterell, Zara L. R., Nicola Beaumont, Matthew Cole, Frances E. Hopkins, Michael Steinke, Richard C. Thompson, and Penelope K. Lindeque. 2020. "Bioavailability of Microplastics to Marine Zooplankton:

Effect of Shape and Infochemicals." *Environmental Science and Technology* 54 (19): 12024–33. https://doi.org/10.1021/acs.est.0c02715.
59. Yang, Yuyi, Wenzhi Liu, Zulin Zhang, Hans-Peter Grossart, and Geoffrey Michael Gadd. 2020. "Microplastics Provide New Microbial Niches in Aquatic Environments." *Applied Microbiology and Biotechnology* 104:6501–11. https://doi.org/10.1007/s00253-020-10704-x.
60. Zettler, Erik R., Tracy J. Mincer, and Linda A. Amaral-Zettler. 2013. "Life in the 'Plastisphere': Microbial Communities on Plastic Marine Debris." *Environmental Science and Technology* 47 (13): 7137–46. https://doi.org/10.1021/es401288x.
61. Amaral-Zettler, Linda A., Erik R. Zettler, Tracy J. Mincer, Michiel A. Klaassen, and Scott M. Gallager. 2021. "Biofouling Impacts on Polyethylene Density and Sinking in Coastal Waters: A Macro/Micro Tipping Point?" *Water Research* 201:117289. https://doi.org/10.1016/j.watres.2021.117289.
62. Amaral-Zettler, Linda A., Erik R. Zettler, and Tracy J. Mincer. 2020. "Ecology of the Plastisphere." *Nature Reviews Microbiology* 18:139–51. https://doi.org/10.1038/s41579-019-0308-0.
63. Bowley, Jake, Craig Baker-Austin, Adam Porter, Rachel Hartnell, and Ceri Lewis. 2021. "Oceanic Hitchhikers: Assessing Pathogen Risks from Marine Microplastic." *Trends in Microbiology* 29 (2): 107–16. https://doi.org/10.1016/j.tim.2020.06.011.
64. Amaral-Zettler, Linda A., Tosca Ballerini, Erik R. Zettler, Alejandro Abdala Asbuna, Alvaro Adamee, Raffaella Casotti, Bruno Dumontet, et al. 2021. "Diversity and Predicted Inter- and Intra-Domain Interactions in the Mediterranean Plastisphere." *Environmental Pollution* 286:117439. https://doi.org/10.1016/j.envpol.2021.117439.
65. Harrison, Jesse P., Timothy J. Hoellein, Melanie Sapp, Alexander S. Tagg, Yon Ju-Nam, and Jesús J. Ojeda. 2018. "Microplastic-Associated Biofilms: A Comparison of Freshwater and Marine Environments." In *Freshwater Microplastics: Emerging Environmental Contaminants?* The Handbook of Environmental Chemistry 58, edited by Martin Wagner and Scott Lambert. https://doi.org/10.1007/978-3-319-61615-5_11.
66. Kooi, Merel, Egbert H. van Nes, Marten Scheffer, and Albert A. Koelmans. 2017. "Ups and Downs in the Ocean: Effects of Biofouling on Vertical Transport of Microplastics." *Environmental Science and Technology* 51 (14). https://doi.org/10.1021/acs.est.6b04702.
67. Kaiser, David, Nicole Kowalski, and Joanna J. Waniek. 2017. "Effects

of Biofouling on the Sinking Behavior of Microplastics." *Environmental Research Letters* 12 (12). http://dx.doi.org/10.1088/1748-9326/aa8e8b.

68. Coyle, Róisín, Gary Hardiman, and Kieran O' Driscoll. 2020. "Microplastics in the Marine Environment: A Review of Their Sources, Distribution Processes, Uptake and Exchange in Ecosystems." *Case Studies in Chemical and Environmental Engineering* 2:100010. https://doi.org/10.1016/j.cscee.2020.100010.
69. Amaral-Zettler, Linda A., Erik R. Zettler, Beth Slikas, Gregory D. Boyd, Donald W. Melvin, Clare E. Morrall, Giora Proskurowski, and Tracy J. Mincer. 2015. "The Biogeography of the Plastisphere: Implications for Policy." *Frontiers in Ecology and the Environment* 13:541–46. https://doi.org/10.1890/150017.
70. Nava, Veronica, and Barbara Leoni. 2021. "A Critical Review of Interactions between Microplastics, Microalgae and Aquatic Ecosystem Function." *Water Research* 188:116476. https://doi.org/10.1016/j.watres.2020.116476.
71. Prata, Joana C., João P. da Costa, Isabel Lopes, Anthony L. Andrady, Armando C. Duarte, and Teresa Rocha-Santos. 2021. "A One Health Perspective of the Impacts of Microplastics on Animal, Human and Environmental Health." *Science of the Total Environment* 777:146094. https://doi.org/10.1016/j.scitotenv.2021.146094.
72. Everaert, Gert, Lisbeth Van Cauwenberghe, Maarten De Rijcke, Albert A. Koelmans, Jan Mees, Michiel Vandegehuchte, and Colin R. Janssen. 2018. "Risk Assessment of Microplastics in the Ocean: Modelling Approach and First Conclusions." *Environmental Pollution* 242 (B):1930–38. https://doi.org/10.1016/j.envpol.2018.07.069.
73. Villarrubia-Gómez, Patricia, Sarah E. Cornell, and Joan Fabres. 2018. "Marine Plastic Pollution as a Planetary Boundary Threat—The Drifting Piece in the Sustainability Puzzle." *Marine Policy* 96:213–20. https://doi.org/10.1016/j.marpol.2017.11.035.
74. Gruber, Nicolas, Dominic Clement, Brendan R. Carter, Richard A. Feely, Steven van Heuven, Mario Hoppema, Masao Ishii, et al. 2019. "The Oceanic Sink for Anthropogenic $CO_2$ from 1994 to 2007." *Science* 363 (6432): 1193–99. https://doi.org/10.1126/science.aau5153.
75. Cole, Matthew, Pennie Lindeque, Elaine Fileman, Claudia Halsband, Rhys Goodhead, Julian Moger, and Tamara S. Galloway. 2013. "Microplastic Ingestion by Zooplankton." *Environmental Science and Technology* 47 (12), 6646–55. https://doi.org/10.1021/es400663f.

76. Setälä, Outi, Vivi Fleming-Lehtinen, and Maiju Lehtiniemi. 2014. "Ingestion and Transfer of Microplastics in the Planktonic Food Web." *Environmental Pollution* 185:77–83. https://doi.org/10.1016/j.envpol.2013.10.013.
77. Sipps, Karli, Georgia Arbuckle-Keil, Robert Chant, Nicole Fahrenfeld, Lori Garzio, Kasey Walsh, and Grace Saba. 2022. "Pervasive Occurrence of Microplastics in Hudson-Raritan Estuary Zooplankton." *Science of the Total Environment* 817:152812. https://doi.org/10.1016/j.scitotenv.2021.152812.
78. Cole, Matthew, Penelope K. Lindeque, Elaine Fileman, James Clark, Ceri Lewis, Claudia Halsband, and Tamara S. Galloway. 2016. "Microplastics Alter the Properties and Sinking Rates of Zooplankton Faecal Pellets." *Environmental Science and Technology* 50 (6): 3239–46. https://doi.org/10.1021/acs.est.5b05905.
79. Pérez-Guevara, Fermín, Priyadarsi D. Roy, Gurusamy Kutralam-Muniasamy, and V. C. Shruti. 2021. "A Central Role for Fecal Matter in the Transport of Microplastics: An Updated Analysis of New Findings and Persisting Questions." *Journal of Hazardous Materials Advances* 4:100021. https://doi.org/10.1016/j.hazadv.2021.100021.
80. Cole, Matthew, Pennie Lindeque, Elaine Fileman, Claudia Halsband, and Tamara S. Galloway. 2015. "The Impact of Polystyrene Microplastics on Feeding, Function and Fecundity in the Marine Copepod *Calanus helgolandicus*." *Environmental Science and Technology* 49 (2): 1130–37. https://doi.org/10.1021/es504525u.
81. Kvale, Karin, A. E. Friederike Prowe, Chia-Te Chien, Landolfi Angela, and Andreas Oschlies. 2020. "The Global Biological Microplastic Particle Sink." *Scientific Reports* 10:16670. https://doi.org/10.1038/s41598-020-72898-4.
82. Gove, Jamison M., Jonathan L. Whitney, Margaret A. McManus, Joey Lecky, Felipe C. Carvalho, Jennifer M. Lynch, Jiwei Li, et al. 2019. "Prey-Size Plastics Are Invading Larval Fish Nurseries." *Proceedings of the National Academy of Sciences* 116 (48): 24143–49. https://doi.org/10.1073/pnas.1907496116.
83. Simon, Matt. 2019. "Baby Fish Feast on Microplastics, and Then Get Eaten." *Wired*. https://www.wired.com/story/baby-fish-are-feasting-on-microplastics/.
84. Nelms, Sarah, James Barnett, Andrew Brownlow, Nick J. Davison, Rob Deaville, Tamara Susan Galloway, P. K. Lindeque, et al. 2019. "Microplastics in Marine Mammals Stranded around the British Coast:

Ubiquitous but Transitory?" *Scientific Reports* 9:1075. https://doi.org/10.1038/s41598-018-37428-3.

85. Lusher, Amy L., Gema Hernandez-Milian, Joanne O'Brien, Simon Berrow, Ian O'Connor, and Rick Officer. 2015. "Microplastic and Macroplastic Ingestion by a Deep Diving, Oceanic Cetacean: The True's Beaked Whale *Mesoplodon mirus*." *Environmental Pollution* 199:185–91. https://doi.org/10.1016/j.envpol.2015.01.023.
86. Carlsson, Pernilla, Cecilie Singdahl-Larsen, and Amy L. Lusher. 2021. "Understanding the Occurrence and Fate of Microplastics in Coastal Arctic Ecosystems: The Case of Surface Waters, Sediments and Walrus (*Odobenus rosmarus*)." *Science of the Total Environment* 792:148308. https://doi.org/10.1016/j.scitotenv.2021.148308; Eriksson, Cecilia, and Harry Burton. 2003. "Origins and Biological Accumulation of Small Plastic Particles in Fur Seals from Macquarie Island." *AMBIO: A Journal of the Human Environment* 32 (6): 380–84. https://doi.org/10.1579/0044-7447-32.6.380.
87. Nelms, Sarah E., Tamara S. Galloway, Brendan J. Godley, Dan S. Jarvis, and Penelope K. Lindeque. 2018. "Investigating Microplastic Trophic Transfer in Marine Top Predators." *Environmental Pollution* 238:999–1007. https://doi.org/10.1016/j.envpol.2018.02.016.
88. National Oceanic and Atmospheric Administration. "Gray Seal." https://www.fisheries.noaa.gov/species/gray-seal.
89. Parry, Wynne. 2010. "Whales Swallow Half a Million Calories in Single Mouthful." *Live Science*. https://www.livescience.com/10332-whales-swallow-million-calories-single-mouthful.html; Fritts, Rachel. 2021. "Baleen Whales Eat Three Times as Much as Scientists Thought." *Science*. https://www.science.org/content/article/baleen-whales-eat-three-times-much-scientists-thought.
90. Garcia-Garin, Odei, Alex Aguilar, Morgana Vighi, Gísli A. Víkingsson, Valérie Chosson, and Asunción Borrell. 2021. "Ingestion of Synthetic Particles by Fin Whales Feeding off Western Iceland in Summer." *Chemosphere* 279:130564. https://doi.org/10.1016/j.chemosphere.2021.130564.
91. Zantis, L. J., T. Bosker, F. Lawler, S. E. Nelms, R. O'Rorke, R. Constantine, M. Sewell, and E. L. Carroll. 2021. "Assessing Microplastic Exposure of Large Marine Filter-Feeders." *Science of the Total Environment* 818:151815. https://doi.org/10.1016/j.scitotenv.2021.151815.
92. Jabeen, Khalida, Bowen Li, Qiqing Chen, Lei Su, Chenxi Wu, Henner Hollert, and Huahong Shi. 2018. "Effects of Virgin Microplastics on

Goldfish (*Carassius auratus*)." *Chemosphere* 213:323–32. https://doi.org/10.1016/j.chemosphere.2018.09.031.

93. Kim, Lia, Sang A. Kim, Tae Hee Kim, Juhea Kim, and Youn-Joo An. 2021. "Synthetic and Natural Microfibers Induce Gut Damage in the Brine Shrimp *Artemia franciscana*." *Aquatic Toxicology* 232:105748. https://doi.org/10.1016/j.aquatox.2021.105748.
94. Carlos de Sá, Luís, Luís G. Luís, and Lúcia Guilhermino. 2015. "Effects of Microplastics on Juveniles of the Common Goby (*Pomatoschistus microps*): Confusion with Prey, Reduction of the Predatory Performance and Efficiency, and Possible Influence of Developmental Conditions." *Environmental Pollution* 196:359–62. https://doi.org/10.1016/j.envpol.2014.10.026.
95. Athey, Samantha N., Samantha D. Albotra, Cessely A. Gordon, Bonnie Monteleone, Pamela Seaton, Anthony L. Andrady, Alison R. Taylor, and Susanne M. Brander. 2020. "Trophic Transfer of Microplastics in an Estuarine Food Chain and the Effects of a Sorbed Legacy Pollutant." *Limnology and Oceanography* 5:154–62. https://doi.org/10.1002/lol2.10130; Baechler, Britta R., Cheyenne D. Stienbarger, Dorothy A. Horn, Jincy Joseph, Alison R. Taylor, Elise F. Granek, and Susanne M. Brander. 2020. "Microplastic Occurrence and Effects in Commercially Harvested North American Finfish and Shellfish: Current Knowledge and Future Directions." *Limnology and Oceanography* 5:113–36. https://doi.org/10.1002/lol2.10122.
96. Duncan, Emily, Annette C. Broderick, Wayne J. Fuller, Tamara S. Galloway, Matthew H. Godfrey, Mark Hamann, Colin J. Limpus, et al. 2019. "Microplastic Ingestion Ubiquitous in Marine Turtles." *Global Change Biology* 25 (2):744–52. https://doi.org/10.1111/gcb.14519.
97. Li, Weixin, Xiaofeng Chen, Minqian Li, Zeming Cai, Han Gong, and Muting Yan. 2022. "Microplastics as an Aquatic Pollutant Affect Gut Microbiota within Aquatic Animals." *Journal of Hazardous Materials* 423 (B): 127094. https://doi.org/10.1016/j.jhazmat.2021.127094.
98. Auta, Helen Shnada, Chijioke Emenike, and Shahul Hamid Fauziah. 2017. "Distribution and Importance of Microplastics in the Marine Environment: A Review of the Sources, Fate, Effects, and Potential Solutions." *Environment International* 102:165–76. https://doi.org/10.1016/j.envint.2017.02.013.
99. Botterell, Zara L. R., Nicola Beaumont, Tarquin Dorrington, Michael Steinke, Richard C. Thompson, and Penelope K. Lindeque. 2019. "Bioavailability and Effects of Microplastics on Marine Zooplankton: A

Review." *Environmental Pollution* 245:98–110. https://doi.org/10.1016/j.envpol.2018.10.065.

100. Nobre, C. R., M. F. M. Santana, A. Malufa, F. S. Cortez, A. Cesar, C. D. S. Pereira, and A. Turra. 2015. "Assessment of Microplastic Toxicity to Embryonic Development of the Sea Urchin *Lytechinus variegatus* (Echinodermata: Echinoidea)." *Marine Pollution Bulletin* 92 (1–2): 99–104. https://doi.org/10.1016/j.marpolbul.2014.12.050.
101. Cormier, Bettie, Chiara Gambardella, Tania Tato, Quentin Perdriat, Elisa Costa, Cloé Veclin, Florane Le Bihanic, et al. 2021. "Chemicals Sorbed to Environmental Microplastics Are Toxic to Early Life Stages of Aquatic Organisms." *Ecotoxicology and Environmental Safety* 208:111665. https://doi.org/10.1016/j.ecoenv.2020.111665.
102. Woods, Madelyn N., Theresa J. Hong, Donaven Baughman, Grace Andrews, David M. Fields, and Patricia A. Matra. 2020. "Accumulation and Effects of Microplastic Fibers in American Lobster Larvae (*Homarus americanus*)." *Marine Pollution Bulletin* 157:111280. https://doi.org/10.1016/j.marpolbul.2020.111280.
103. Jensen, Lene H., Cherie A. Motti, Anders L. Garm, Hemerson Tonin, and Frederieke J. Kroon. 2019. "Sources, Distribution and Fate of Microfibres on the Great Barrier Reef, Australia." *Scientific Reports* 9:9021. https://doi.org/10.1038/s41598-019-45340-7.
104. Utami, Dwi Amanda, Lars Reuning, Olga Konechnaya, and Jan Schwarzbauer. 2021. "Microplastics as a Sedimentary Component in Reef Systems: A Case Study from the Java Sea." *Sedimentology* 68:2270–92. https://doi.org/10.1111/sed.12879.
105. Chapron, Leila, Erwan Peru, A. Engler, Jean-Francois Ghiglione, Anne-Leila Meistertzheim, Audry M. Pruski, Autun Purser, et al. 2018. "Macro- and Microplastics Affect Cold-Water Corals Growth, Feeding and Behaviour." *Scientific Reports* 8:15299. https://doi.org/10.1038/s41598-018-33683-6.
106. John, Juliana, A. R. Nandhini, Padmanaban Velayudhaperumal Chellam, and Mika Sillanpää. 2021. "Microplastics in Mangroves and Coral Reef Ecosystems: A Review." *Environmental Chemistry Letters*. https://doi.org/10.1007/s10311-021-01326-4.
107. Montano, Simone, Davide Seveso, Davide Maggioni, Paolo Galli, Stefano Corsarini, and Francesco Saliu. 2020. "Spatial Variability of Phthalates Contamination in the Reef-Building Corals *Porites lutea, Pocillopora verrucosa* and *Pavona varians*." *Marine Pollution Bulletin* 155:111117. https://doi.org/10.1016/j.marpolbul.2020.111117.

108. Reichert, Jessica, Angelina L. Arnold, Nils Hammer, Ingo B. Miller, Marvin Rades, Patrick Schubert, Maren Ziegler, and Thomas Wilke. 2021. "Reef-Building Corals Act as Long-Term Sink for Microplastic." *Global Change Biology* 28 (1): 33–45. https://doi.org/10.1111/gcb.15920.
109. Foley, Carolyn J., Zachary S. Feiner, Timothy D. Malinich, and Tomas O. Höök. 2018. "A Meta-Analysis of the Effects of Exposure to Microplastics on Fish and Aquatic Invertebrates." *Science of the Total Environment* 631–32:550–59. https://doi.org/10.1016/j.scitotenv.2018.03.046.
110. Hou, Loren, Caleb D. McMahan, Rae E. McNeish, Keenan Munno, Chelsea M. Rochman, and Timothy J. Hoellein. 2021. "A Fish Tale: A Century of Museum Specimens Reveal Increasing Microplastic Concentrations in Freshwater Fish." *Ecological Applications* 31 (5). https://doi.org/10.1002/eap.2320.
111. Savoca, Matthew S., Alexandra G. McInturf, and Elliott L. Hazen. 2021. "Plastic Ingestion by Marine Fish Is Widespread and Increasing." *Global Change Biology* 27 (10): 2188–99. https://doi.org/10.1111/gcb.15533.
112. Collard, France, Bernard Gilbert, Philippe Compère, Gauthier Eppe, Krishna Das, Thierry Jauniaux, and Eric Parmentier. 2017. "Microplastics in Livers of European Anchovies (*Engraulis encrasicolus, L.*)" *Environmental Pollution* 229:1000–1005. https://doi.org/10.1016/j.envpol.2017.07.089; McIlwraith, Hayley K., Joel Kim, Paul Helm, Satyendra P. Bhavsar, Jeremy S. Metzger, and Chelsea M. Rochman. 2021. "Evidence of Microplastic Translocation in Wild-Caught Fish and Implications for Microplastic Accumulation Dynamics in Food Webs." *Environmental Science and Technology* 55 (18): 12372–82. https://doi.org/10.1021/acs.est.1c02922; Sequeira, Inês F., Joana C. Prata, João P. da Costa, Armando C. Duarte, and Teresa Rocha-Santos. 2020. "Worldwide Contamination of Fish with Microplastics: A Brief Global Overview." *Marine Pollution Bulletin* 160:111681. https://doi.org/10.1016/j.marpolbul.2020.111681.
113. De Sales-Ribeiro, Carolina, Yeray Brito-Casillas, and Antonio Fernandez. 2020. "An End to the Controversy over the Microscopic Detection and Effects of Pristine Microplastics in Fish Organs." *Scientific Reports* 10:12434. https://doi.org/10.1038/s41598-020-69062-3; Zeytin, Sinem, Gretchen Wagner, Nick Mackay-Roberts, Gunnar Gerdts, Erwin Schuirmann, Sven Klockmann, and Matthew Slater. 2020. "Quantifying Microplastic Translocation from Feed to the Fillet in European

Sea Bass *Dicentrarchus labrax*." *Marine Pollution Bulletin* 156:111210. https://doi.org/10.1016/j.marpolbul.2020.111210.

114. Ribeiro, Francisca, Elvis D. Okoffo, Jake W. O'Brien, Sarah Fraissinet-Tachet, Stacey O'Brien, Michael Gallen, Saer Samanipour, et al. 2020. "Quantitative Analysis of Selected Plastics in High-Commercial-Value Australian Seafood by Pyrolysis Gas Chromatography Mass Spectrometry." *Environmental Science and Technology* 54 (15), 9408–17. https://doi.org/10.1021/acs.est.0c02337.
115. Dawson, Amanda L., Marina F. M. Santana, Michaela E. Miller, and Frederieke J. Kroon. 2021. "Relevance and Reliability of Evidence for Microplastic Contamination in Seafood: A Critical Review Using Australian Consumption Patterns as a Case Study." *Environmental Pollution* 276:116684. https://doi.org/10.1016/j.envpol.2021.116684.
116. Mattsson, Karin, Elyse V. Johnson, Anders Malmendal, Sara Linse, Lars-Anders Hansson, and Tommy Cedervall. 2017. "Brain Damage and Behavioural Disorders in Fish Induced by Plastic Nanoparticles Delivered through the Food Chain." *Scientific Reports* 7:11452. https://doi.org/10.1038/s41598-017-10813-0.
117. Cunningham, Eoghan M., Amy Mundye, Louise Kregting, Jaimie T. A. Dick, Andrew Crump, Gillian Riddell, and Gareth Arnott. 2021. "Animal Contests and Microplastics: Evidence of Disrupted Behaviour in Hermit Crabs *Pagurus bernhardus*." *Royal Society Open Science* 8 (10). http://doi.org/10.1098/rsos.211089.
118. Chae, Yooeun, Dokyung Kim, Shin Woong Kim, and Youn-Joo An. 2018. "Trophic Transfer and Individual Impact of Nano-Sized Polystyrene in a Four-Species Freshwater Food Chain." *Scientific Reports* 8:284. https://doi.org/10.1038/s41598-017-18849-y.
119. Dawson, Amanda, So Kawaguchi, Catherine K. King, Kathy A. Townsend, Robert King, Wilhelmina M. Huston, and Susan M. Bengtson Nash. 2018. "Turning Microplastics into Nanoplastics through Digestive Fragmentation by Antarctic Krill." *Nature Communications* 9:1001. https://doi.org/10.1038/s41467-018-03465-9.
120. Seo, Hannah. 2021. "Fish Farming Has a Plastic Problem." *Environmental Health News*. https://www.ehn.org/plastic-in-farmed-fish-2650268080.html.
121. FAO (Food and Agriculture Organization). 2020. *The State of World Fisheries and Aquaculture 2020: Sustainability in Action*. Rome: FAO. https://doi.org/10.4060/ca9229en.
122. Thiele, Christina J., Malcolm D. Hudson, Andrea E. Russell, Marilin

Saluveer, and Giovanna Sidaoui-Haddad. 2021. "Microplastics in Fish and Fishmeal: An Emerging Environmental Challenge?" *Scientific Reports* 11:2045. https://doi.org/10.1038/s41598-021-81499-8.

123. Hanachi, Parichehr, Samaneh Karbalaei, Tony R. Walker, Matthew Cole, and Seyed V. Hosseini. 2019. "Abundance and Properties of Microplastics Found in Commercial Fish Meal and Cultured Common Carp (*Cyprinus carpio*)." *Environmental Science and Pollution Research* 26:23777–87. https://doi.org/10.1007/s11356-019-05637-6.
124. Mathalon, Alysse, and Paul Hill. 2014. "Microplastic Fibers in the Intertidal Ecosystem Surrounding Halifax Harbor, Nova Scotia." *Marine Pollution Bulletin* 81 (1): 69–79. https://doi.org/10.1016/j.marpolbul.2014.02.018.
125. Van Cauwenberghe, Lisbeth, and Colin R. Janssen. 2014. "Microplastics in Bivalves Cultured for Human Consumption." *Environmental Pollution* 193:65–70. https://doi.org/10.1016/j.envpol.2014.06.010.
126. Akoueson, Fleurine, Lisa M. Sheldon, Evangelos Danopoulos, Steve Morris, Jessica Hotten, Emma Chapman, Jiana Li, and Jeanette M. Rotchell. 2020. "A Preliminary Analysis of Microplastics in Edible versus Non-Edible Tissues from Seafood Samples." *Environmental Pollution* 263 (A): 114452. https://doi.org/10.1016/j.envpol.2020.114452.

**Chapter 3: A Land Corrupted**

1. Tian, Zhenyu, Haoqi Zhao, Katherine T. Peter, Melissa Gonzalez, Jill Wetzel, Christopher Wu, Ximin Hu, et al. 2021. "A Ubiquitous Tire Rubber–Derived Chemical Induces Acute Mortality in Coho Salmon." *Science* 371 (6525): 185–89. https://doi.org/10.1126/science.abd6951.
2. American Chemical Society National Historic Chemical Landmarks. n.d. "U.S. Synthetic Rubber Program." http://www.acs.org/content/acs/en/education/whatischemistry/landmarks/syntheticrubber.html.
3. Wagner, Stephan, Thorsten Hüffer, Philipp Klöckner, Maren Wehrhahn, Thilo Hofmann, and Thorsten Reemtsma. 2018. "Tire Wear Particles in the Aquatic Environment: A Review on Generation, Analysis, Occurrence, Fate and Effects." *Water Research* 139:83–100. https://doi.org/10.1016/j.watres.2018.03.051.
4. Tamis, Jacqueline E., Albert A. Koelmans, Rianne Dröge, Nicolaas H. B. M. Kaag, Marinus C. Keur, Peter C. Tromp, and Ruud H. Jongbloed. 2021. "Environmental Risks of Car Tire Microplastic Particles and Other Road Runoff Pollutants." *Microplastics and Nanoplastics* 1:10. https://doi.org/10.1186/s43591-021-00008-w.

5. Huntink, Nicolaas Maria. 2003. "Durability of Rubber Products: Development of New Antidegradants for Long-Term Protection." PhD thesis, Twente University Press. https://research.utwente.nl/en/publications/durability-of-rubber-products-development-of-new-antidegradants-f.
6. Csere, Csaba. 2000. "A Look Behind the Tire Hysteria." *Car and Driver*. https://www.caranddriver.com/features/a15139795/a-look-behind-the-tire-hysteria/.
7. Kole, Pieter Jan, Ansje J. Löhr, Frank G. A. J. Van Belleghem, and Ad M. J. Ragas. 2017. "Wear and Tear of Tyres: A Stealthy Source of Microplastics in the Environment." *International Journal of Environmental Research and Public Health* 14 (10): 1265. https://dx.doi.org/10.3390%2Fijerph14101265.
8. McIntyre, Jenifer K., Jasmine Prat, James Cameron, Jillian Wetzel, Emma Mudrock, Katherine T. Peter, Zhenyu Tian, et al. 2021. "Treading Water: Tire Wear Particle Leachate Recreates an Urban Runoff Mortality Syndrome in Coho but Not Chum Salmon." *Environmental Science and Technology* 55 (17): 11767–74. https://doi.org/10.1021/acs.est.1c03569.
9. Brinkmann, Markus, David Montgomery, Summer Selinger, Justin G. P. Miller, Eric Stock, Alper James Alcaraz, Jonathan K. Challis, et al. 2022. "Acute Toxicity of the Tire Rubber-Derived Chemical 6PPD-Quinone to Four Fishes of Commercial, Cultural, and Ecological Importance." *Environmental Science and Technology Letters*. https://doi.org/10.1021/acs.estlett.2c00050.
10. Simon, Matt. 2020. "This Bizarre Insect Is Building Shelters Out of Microplastic." *Wired*. https://www.wired.com/story/caddisfly-microplastic/.
11. Triebskorn, Rita, Thomas Braunbeck, Tamara Grummt, Lisa Hanslik, Sven Huppertsberg, Martin Jekel, Thomas P. Knepper, et al. 2019. "Relevance of Nano- and Microplastics for Freshwater Ecosystems: A Critical Review." *TrAC Trends in Analytical Chemistry* 110:375–92. https://doi.org/10.1016/j.trac.2018.11.023.
12. Hurley, Rachel, Jamie Woodward, and James J. Rothwell. 2018. "Microplastic Contamination of River Beds Significantly Reduced by Catchment-Wide Flooding." *Nature Geoscience* 11:251–57. https://doi.org/10.1038/s41561-018-0080-1.
13. Leslie, H. A., Sicco H. Brandsma, M. J. M. Van Velzen, and Andre Dick Vethaak. 2017. "Microplastics en Route: Field Measurements in the Dutch River Delta and Amsterdam Canals, Wastewater Treatment Plants, North Sea Sediments and Biota." *Environment International* 101:133–42. https://doi.org/10.1016/j.envint.2017.01.018.

14. Moore, Charles James, Gwendolyn L. Lattin, and A. F. Zellers. 2011. "Quantity and Type of Plastic Debris Flowing from Two Urban Rivers to Coastal Waters and Beaches of Southern California." *Journal of Integrated Coastal Zone Management* 11 (1): 65–73. http://dx.doi.org/10.5894/rgci194.
15. Napper, Imogen E., Anju Baroth, Aaron C. Barrett, Sunanda Bhola, Gawsia W. Chowdhury, Bede F. R. Davies, Emily M. Duncan, et al. 2021. "The Abundance and Characteristics of Microplastics in Surface Water in the Transboundary Ganges River." *Environmental Pollution* 274:116348. https://doi.org/10.1016/j.envpol.2020.116348.
16. Waterfront Partnership of Baltimore. n.d. "Mr. Trash Wheel: A Proven Solution to Ocean Plastics." https://www.mrtrashwheel.com/.
17. Simon, Matt. 2019. "7 Trillion Microplastic Particles Pollute the San Francisco Bay Each Year." *Wired*. https://www.wired.com/story/microplastic-san-francisco-bay/.
18. Napper, Imogen E., Bede F. R. Davies, Heather Clifford, Sandra Elvin, Heather J. Koldewey, Paul A. Mayewski, Kimberley R. Miner, et al. 2020. "Reaching New Heights in Plastic Pollution: Preliminary Findings of Microplastics on Mount Everest." *One Earth* 3 (5): 621–30. https://doi.org/10.1016/j.oneear.2020.10.020.
19. Green, Dannielle S., Andrew D. W. Tongue, and Bas Boots. 2021. "The Ecological Impacts of Discarded Cigarette Butts." *Trends in Ecology and Evolution* 37 (2): 183–92. https://doi.org/10.1016/j.tree.2021.10.001.
20. Belzagui, Francisco, Valentina Buscio, Carmen Gutiérrez-Bouzán, and Mercedes Vilaseca. 2021. "Cigarette Butts as a Microfiber Source with a Microplastic Level of Concern." *Science of the Total Environment* 762:144165. https://doi.org/10.1016/j.scitotenv.2020.144165.
21. Ocean Conservancy. 2017. *International Coastal Cleanup Report*. Washington, DC: Ocean Conservancy. https://oceanconservancy.org/wp-content/uploads/2017/06/International-Coastal-Cleanup_2017-Report.pdf.
22. Parson, Ann. 2021. "How Paving with Plastic Could Make a Dent in the Global Waste Problem." Yale Environment 360. https://e360.yale.edu/features/how-paving-with-plastic-could-make-a-dent-in-the-global-waste-problem.
23. Magnusson, Kerstin, Karin Eliasson, Anna Fråne, Kalle Haikonen, Johan Hultén, Mikael Olshammar, Johanna Stadmark, and Anais Voisin. 2016. *Swedish Sources and Pathways for Microplastics to the Marine Environment*. Stockholm: IVL Swedish Environmental Research Institute. https://www.ccb.se/documents/ML_background/SE_Study_MP_sources.pdf.

24. Werbowski, Larissa M., Alicia N. Gilbreath, Keenan Munno, Xia Zhu, Jelena Grbic, Tina Wu, Rebecca Sutton, et al. 2021. “Urban Stormwater Runoff: A Major Pathway for Anthropogenic Particles, Black Rubbery Fragments, and Other Types of Microplastics to Urban Receiving Waters.” *ACS ES&T Water* 1 (6): 1420–28. https://doi.org/10.1021/acsest water.1c00017.
25. Fidra. 2020. “Microplastic Loss from Artificial (3G) Pitches in Context of the ECHA Proposed Restriction of Microplastics Intentionally Added to Products.” https://www.plasticsoupfoundation.org/wp-content/up loads/2020/09/Fidra-Microplastic-loss-from-artificial-3G-pitches_v2 -.pdf.
26. Løkkegaard, Hanne, Bjørn Malmgren-Hansen, and Nils H. Nilsson. 2018. *Mass Balance of Rubber Granulate Lost from Artificial Turf Fields, Focusing on Discharge to the Aquatic Environment*. Viborg: Danish Technological Institute. https://www.genan.eu/wp-content/up loads/2020/02/Teknologisk-Institut_Mass-balance-of-rubber-granu late-lost-from-artificial-turf-fields_May-2019_v1.pdf.
27. Hann, Simon, Chris Sherrington, Olly Jamieson, Molly Hickman, and Ayesha Bapasola. 2018. “Investigating Options for Reducing Releases in the Aquatic Environment of Microplastics Emitted by Products.” Eunomia. https://www.eunomia.co.uk/reports-tools/investigating-op tions-for-reducing-releases-in-the-aquatic-environment-of-microplas tics-emitted-by-products/.
28. Swedish Environmental Protection Agency. 2021. “Microplastics in the Environment 2019: Report on a Government Commission.” Bromma: Swedish EPA. https://www.naturvardsverket.se/globalassets/media/pub likationer-pdf/6900/978-91-620-6957-5.pdf.
29. Gilbreath, Alicia, Lester McKee, Ila Shimabuku, Diana Lin, Larissa M. Werbowski, Xia Zhu, Jelena Grbic, and Chelsea Rochman. 2019. “Multiyear Water Quality Performance and Mass Accumulation of PCBs, Mercury, Methylmercury, Copper, and Microplastics in a Bioretention Rain Garden.” *Journal of Sustainable Water in the Built Environment* 5 (4). https://doi.org/10.1061/JSWBAY.0000883; Rochman, Chelsea M., Keenan Munno, Carolynn Box, Anna Cummins, Xia Zhu, and Rebecca Sutton. 2021. “Think Global, Act Local: Local Knowledge Is Critical to Inform Positive Change When It Comes to Microplastics.” *Environmental Science and Technology* 55 (1): 4–6. https://doi.org/10 .1021/acs. est.0c05746.
30. Moran, Kelly, Ezra Miller, Miguel Mendez, Shelly Moore, Alicia Gilbreath, Rebecca Sutton, and Diana Lin. 2021. *A Synthesis of Microplastic*

*Sources and Pathways to Urban Runoff.* SFEI Technical Report: SFEI Contribution #1049. Richmond, CA: San Francisco Estuary Institute.

31. Simon, Matt. 2021. "Climate Change Is Turning Cities into Ovens." *Wired.* https://www.wired.com/story/climate-change-is-turning-cities-into-ovens/.
32. United Nations. 2018. "68% of the World Population Projected to Live in Urban Areas by 2050, Says UN." https://www.un.org/development/desa/en/news/population/2018-revision-of-world-urbanization-prospects.html.
33. Anderson, Kendall W. 2019. *Occupational Health Risks from Class B Biosolids.* Chicago: Illinois Injury Prevention Center. http://illinoisinjuryprevention.org/Industry_Spotlight_Class%20B%20Biosolids%20Brief.pdf.
34. Nizzetto, Luca, Sindre Langaas, and Martyn Futter. 2016. "Pollution: Do Microplastics Spill on to farm soils?" *Nature* 537 (488). https://doi.org/10.1038/537488b.
35. Nizzetto, Luca, Martyn Futter, and Sindre Langaas. 2016. "Are Agricultural Soils Dumps for Microplastics of Urban Origin?" *Environmental Science and Technology* 50 (20): 10777–79. https://doi.org/10.1021/acs.est.6b04140.
36. Lusher, Amy L., Rachel Hurley, Christian Vogelsang, Luca Nizzetto, and Marianne Olsen. 2017. "Mapping Microplastics in Sludge." Norwegian Institute for Water Research. https://niva.brage.unit.no/niva-xmlui/handle/11250/2493527.
37. Mahon, Anne Marie, Brendan O'Connell, Mark Gerard Healy, Ian O'Connor, Rick Andrew Officer, Roisin Nash, and Liam Morrison. 2017. "Microplastics in Sewage Sludge: Effects of Treatment." *Environmental Science and Technology* 51 (2): 810–18. http://dx.doi.org/10.1021/acs.est.6b04048.
38. Magnusson, Kerstin, and Fredrik Norén. 2014. *Screening of Microplastic Particles in and Down-Stream of a Wastewater Treatment Plant.* Stockholm: IVL Swedish Environmental Research Institute. https://www.diva-portal.org/smash/get/diva2:773505/FULLTEXT01.pdf.
39. Mintenig, S. M., I. Int-Veen, M. G. J. Löder, S. Primpke, and G. Gerdts. 2017. "Identification of Microplastic in Effluents of Waste Water Treatment Plants Using Focal Plane Array-Based Micro-Fourier-Transform Infrared Imaging." *Water Research* 108:365–72. https://doi.org/10.1016/j.watres.2016.11.015.
40. Sujathan, Surya, Ann-Kathrin Kniggendorf, Arun Kumar, Bernhard

Roth, Karl-Heinz Rosenwinkel, and Regina Nogueira. 2017. "Heat and Bleach: A Cost-Efficient Method for Extracting Microplastics from Return Activated Sludge." *Archives of Environmental Contamination and Toxicology* 73:641–48. https://doi.org/10.1007/s00244-017-0415-8.

41. Gavigan, Jenna, Timnit Kefela, Ilan Macadam-Somer, Sangwon Suh, and Roland Geyer. 2020. "Synthetic Microfiber Emissions to Land Rival Those to Waterbodies and Are Growing." *PLoS ONE* 15 (9): e0237839. https://doi.org/10.1371/journal.pone.0237839; Gammon, Katherine. 2020. "Groundbreaking Study Finds 13.3 Quadrillion Plastic Fibers in California's Environment." *Guardian.* https://www.theguardian.com/us-news/2020/oct/16/plastic-waste-microfibers-california-study.
42. Zubris, Kimberly Ann V., and Brian K. Richards. 2005. "Synthetic Fibers as an Indicator of Land Application of Sludge." *Environmental Pollution* 138 (2): 201–11. https://doi.org/10.1016/j.envpol.2005.04.013.
43. Corradini, Fabio, Pablo Meza, Raúl Eguiluz, Francisco Casado, Esperanza Huerta-Lwanga, and Violette Geissen. 2019. "Evidence of Microplastic Accumulation in Agricultural Soils from Sewage Sludge Disposal." *Science of the Total Environment* 671:411–20. https://doi.org/10.1016/j.scitotenv.2019.03.368.
44. Crossman, Jill, Rachel R. Hurley, Martyn Futter, and Luca Nizzetto. 2020. "Transfer and Transport of Microplastics from Biosolids to Agricultural Soils and the Wider Environment." *Science of the Total Environment* 724:138334. https://doi.org/10.1016/j.scitotenv.2020.138334.
45. Roebroek, Caspar T. J., Shaun Harrigan, Tim H. M. van Emmerik, Calum Baugh, Dirk Eilander, Christel Prudhomme, and Florian Pappenberger. 2021. "Plastic in Global Rivers: Are Floods Making It Worse?" *Environmental Research Letters* 16 (2). http://dx.doi.org/10.1088/1748-9326/abd5df.
46. Weithmann, Nicolas, Julia N. Möller, Martin G. J. Löder, Sarah Piehl, Christian Laforsch, and Ruth Freitag. 2018. "Organic Fertilizer as a Vehicle for the Entry of Microplastic into the Environment." *Science Advances* 4 (4). https://doi.org/10.1126/sciadv.aap8060.
47. Kelly, John J., Maxwell G. London, Amanda R. McCormick, Miguel Rojas, John W. Scott, and Timothy J. Hoellein. 2021. "Wastewater Treatment Alters Microbial Colonization of Microplastics." *PLoS ONE* 16 (1): e0244443. https://doi.org/10.1371/journal.pone.0244443.
48. McCormick, Amanda R., Timothy J. Hoellein, Maxwell G. London, Joshua Hittie, John W. Scott, and John J. Kelly. 2016. "Microplastic in Surface Waters of Urban Rivers: Concentration, Sources, and Associated

Bacterial Assemblages." *Ecosphere* 7 (11). https://doi.org/10.1002/ecs2.1556.

49. Pham, Dung Ngoc, Lerone Clark, and Mengyan Li. 2021. "Microplastics as Hubs Enriching Antibiotic-Resistant Bacteria and Pathogens in Municipal Activated Sludge." *Journal of Hazardous Materials Letters* 2:100014. https://doi.org/10.1016/j.hazl.2021.100014.
50. New Jersey Institute of Technology. 2021. "How Our Microplastic Waste Becomes 'Hubs' for Pathogens, Antibiotic-Resistant Bacteria." https://www.eurekalert.org/pub_releases/2021-03/njio-hom031921.php.
51. Arias-Andres, Maria, Uli Klümper, Keilor Rojas-Jimenez, and Hans-Peter Grossart. 2018. "Microplastic Pollution Increases Gene Exchange in Aquatic Ecosystems." *Environmental Pollution* 237:253–61. https://doi.org/10.1016/j.envpol.2018.02.058.
52. Landis, Thomas D., and R. Kasten Dumroese. 2009. "Using Polymer-Coated Controlled-Release Fertilizers in the Nursery and after Outplanting." *Forest Nursery Notes* (Winter): 5–12. https://www.srs.fs.usda.gov/pubs/34172.
53. International Plant Nutrition Institute. n.d. *Coated Fertilizer*. Norcross, GA: INPNI. http://www.ipni.net/publication/nss.nsf/0/33C6A283CC38EE26852579AF007682E3/$FILE/NSS-20%20Coated%20Fertilizer.pdf.
54. Katsumi, Naoya, Takasei Kusube, Seiya Nagao, and Hiroshi Okochi. 2021. "Accumulation of Microcapsules Derived from Coated Fertilizer in Paddy Fields." *Chemosphere* 267:129185. https://doi.org/10.1016/j.chemosphere.2020.129185.
55. Katsumi, Naoya, Takasei Kusube, Seiya Nagao, and Hiroshi Okochi. 2020. "The Role of Coated Fertilizer Used in Paddy Fields as a Source of Microplastics in the Marine Environment." *Marine Pollution Bulletin* 161 (B): 111727. https://doi.org/10.1016/j.marpolbul.2020.111727.
56. Spence, Carol Lea. 2007. "'Revolutionary' Greenhouse Has Roots at UK." University of Kentucky College of Agriculture, Food, and Environment. https://news.ca.uky.edu/article/'revolutionary'-greenhouse-has-roots-uk.
57. Grubinger, Vern. 2004. "Plastic Mulch Primer." University of Vermont Extension. https://www.uvm.edu/vtvegandberry/factsheets/plasticprimer.html.
58. Kasirajan, Subrahmaniyan, and Mathieu Ngouajio. 2012. "Polyethylene and Biodegradable Mulches for Agricultural Applications: A Review." *Agronomy for Sustainable Development* 32:501–29. http://dx.doi.org/10.1007/s13593-011-0068-3.

59. Maughan, Tiffany, and Dan Drost. 2016. "Use of Plastic Mulch for Vegetable Production." Utah State University Extension. https://extension.usu.edu/productionhort/files-ou/Use-of-Plastic-Mulch-for-Vegetable-Production.pdf.
60. Shrefler, Jim, and Lynn Brandenberger. 2016. "Use of Plastic Mulch and Row Covers in Vegetable Production." Oklahoma State University Extension. https://extension.okstate.edu/fact-sheets/use-of-plastic-mulch-and-row-covers-in-vegetable-production.html; Wang, Ting, Yini Ma, and Rong Ji. 2020. "Aging Processes of Polyethylene Mulch Films and Preparation of Microplastics with Environmental Characteristics." *Bulletin of Environmental Contamination and Toxicology* 107:736–40. https://doi.org/10.1007/s00128-020-02975-x.
61. Steinmetz, Zacharias, Claudia Wollmann, Miriam Schaefer, Christian Buchmann, Jan David, Josephine Tröger, Katherine Muñoz, Oliver Frör, and Gabriele Ellen Schaumann. 2016. "Plastic Mulching in Agriculture. Trading Short-Term Agronomic Benefits for Long-Term Soil Degradation?" *Science of the Total Environment* 550:690–705. https://doi.org/10.1016/j.scitotenv.2016.01.153.
62. Moore, Jenny, and Annette Wszelaki. 2016. "Plastic Mulch in Fruit and Vegetable Production: Challenges for Disposal." Report No. FA-2016-02. https://ag.tennessee.edu/biodegradablemulch/Documents/Plastic_Mulch_in_Fruit_and_Vegetable_Production_12_20factsheet.pdf.
63. Sintim, Henry Y., and Markus Flury. 2017. "Is Biodegradable Plastic Mulch the Solution to Agriculture's Plastic Problem?" *Environmental Science and Technology* 51 (3): 1068–69. https://doi.org/10.1021/acs.est.6b06042.
64. Beriot, Nicolas, Joost Peek, Raul Zornoza, Violette Geissen, and Esperanza Huerta Lwanga. 2021. "Low Density-Microplastics Detected in Sheep Faeces and Soil: A Case Study from the Intensive Vegetable Farming in Southeast Spain." *Science of the Total Environment* 755 (1): 142653. https://doi.org/10.1016/j.scitotenv.2020.142653.
65. Wenqing, He, Liu Enke, Liu Qin, Liu Shuang, Neil C. Turner, and Yan Changrong. 2014. "Plastic-Film Mulch in Chinese Agriculture: Importance and Problems." *World Agriculture* 4 (2): 32–36.
66. Neretin, Lev. 2021. "Fields of Plastics." International Institute for Sustainable Development. http://sdg.iisd.org/commentary/guest-articles/fields-of-plastics/.
67. Guo, Jing-Jie, Xian-Pei Huang, Lei Xiang, Yi-Ze Wang, Yan-Wen Li, Hui Li, Quan-Ying Cai, Ce-Hui Mo, and Ming-Hung Wong. 2020. "Source, Migration and Toxicology of Microplastics in Soil."

*Environment International* 137:105263. https://doi.org/10.1016/j.envint.2019.105263.

68. Huang, Yi, Qin Liu, Weiqian Jia, Changrong Yan, and Jie Wang. "Agricultural Plastic Mulching as a Source of Microplastics in the Terrestrial Environment." *Environmental Pollution* 260:114096. https://doi.org/10.1016/j.envpol.2020.114096.
69. Zhang, Huan, Markus Flury, Carol Miles, Hang Liu, and Lisa DeVetter. 2020. "Soil-Biodegradable Plastic Mulches Undergo Minimal in-Soil Degradation in a Perennial Raspberry System after 18 Months." *Horticulturae* 6 (3): 47. https://doi.org/10.3390/horticulturae6030047.
70. Zhang, G. S., and Y. F. Liu. 2018. "The Distribution of Microplastics in Soil Aggregate Fractions in Southwestern China." *Science of the Total Environment* 642:12–20. https://doi.org/10.1016/j.scitotenv.2018.06.004.
71. Zhang, Dan, Ee Ling Ng, Wanli Hu, Hongyuan Wang, Pablo Galaviz, Hude Yang, Wentao Sun, et al. 2020. "Plastic Pollution in Croplands Threatens Long-Term Food Security." *Global Change Biology* 26:3356–67. https://doi.org/10.1111/gcb.15043.
72. Caparrós-Martínez, José Luis, Nuria Rueda-Lópe, Juan Milán-García, and Jaime de Pablo Valenciano. 2020. "Public Policies for Sustainability and Water Security: The Case of Almeria (Spain)." *Global Ecology and Conservation* 23:e01037. https://doi.org/10.1016/j.gecco.2020.e01037.
73. Dahl, Martin, Sanne Bergman, Mats Björk, Elena Diaz-Almela, Maria Granberg, Martin Gullström, Carmen Leiva-Dueñas, et al. 2021. "A Temporal Record of Microplastic Pollution in Mediterranean Seagrass Soils." *Environmental Pollution* 273:116451. https://doi.org/10.1016/j.envpol.2021.116451.
74. FAO (Food and Agriculture Organization). 2021. *Assessment of Agricultural Plastics and Their Sustainability: A Call for Action*. Rome: FAO. https://doi.org/10.4060/cb7856en.
75. Rillig, Matthias C., Lisa Ziersch, and Stefan Hempel. 2017. "Microplastic Transport in Soil by Earthworms." *Scientific Reports* 7:1362. http://dx.doi.org/10.1038/s41598-017-01594-7.
76. Lwanga, Esperanza Huerta, Hennie Gertsen, Harm Gooren, Piet Peters, Tamás Salánki, Martine van der Ploeg, Ellen Besseling, Albert A. Koelmans, and Violette Geissen. 2017. "Incorporation of Microplastics from Litter into Burrows of *Lumbricus terrestris*." *Environmental Pollution* 220:523–31. https://doi.org/10.1016/j.envpol.2016.09.096.
77. Huerta Lwanga, Esperanza, Hennie Gertsen, Harm Gooren, Piet Peters, Tamás Salánki, Martine van der Ploeg, Ellen Besseling, et al. 2016.

"Microplastics in the Terrestrial Ecosystem: Implications for Lumbricus terrestris (*Oligochaeta, Lumbricidae*)." *Environmental Science and Technology* 50 (5): 2685–91. https://doi.org/10.1021/acs.est.5b05478.

78. Rillig, Matthias C., Rosolino Ingraffia, and Anderson A. de Souza Machado. 2017. "Microplastic Incorporation into Soil in Agroecosystems." *Frontiers in Plant Science* 8:1805. https://doi.org/10.3389/fpls.2017.01805.
79. O'Connor, David, Shizhen Pan, Zhengtao Shen, Yinan Song, Yuanliang Jin, Wei-Min Wu, and Deyi Hou. 2019. "Microplastics Undergo Accelerated Vertical Migration in Sand Soil Due to Small Size and Wet-Dry Cycles." *Environmental Pollution* 249:527–34. https://doi.org/10.1016/j.envpol.2019.03.092.
80. De Souza Machado, Anderson Abel, Chung Wai Lau, Jennifer Till, Werner Kloas, Anika Lehmann, Roland Becker, and Matthias C. Rillig. 2018. "Impacts of Microplastics on the Soil Biophysical Environment." *Environmental Science and Technology* 52 (17): 9656–65. https://doi.org/10.1021/acs.est.8b02212.
81. Lili, Lei, Siyu Wu, Shibo Lu, Mengting Liu, Yang Song, Zhenhuan Fu, Huahong Shi, Kathleen M. Raley-Susman, and Defu He. 2018. "Microplastic Particles Cause Intestinal Damage and Other Adverse Effects in Zebrafish *Danio Rerio* and Nematode *Caenorhabditis Elegans*." *Science of the Total Environment*. 619–620:1–8. https://doi.org/10.1016/j.scitotenv.2017.11.103.
82. Song, Yang, Chengjin Cao, Rong Qiu, Jiani Hu, Mengting Liu, Shibo Lu, Huahong Shi, Kathleen M. Raley-Susman, and Defu He. 2019. "Uptake and Adverse Effects of Polyethylene Terephthalate Microplastics Fibers on Terrestrial Snails (*Achatina fulica*) after Soil Exposure." *Environmental Pollution* 250:447–55. https://doi.org/10.1016/j.envpol.2019.04.066.
83. Selonen, Salla, Andraž Dolar, Anita Jemec Kokalj, Tina Skalar, Lidia Parramon Dolcet, Rachel Hurley, and Cornelis A. M. van Gestel. 2020. "Exploring the Impacts of Plastics in Soil: The Effects of Polyester Textile Fibers on Soil Invertebrates." *Science of the Total Environment* 700:134451. https://doi.org/10.1016/j.scitotenv.2019.134451; Zhang, Shaoliang, Jiuqi Wang, Pengke Yan, Xinhua Hao, Bing Xu, Wan Wang, and Muhammad Aurangzeib. 2021. "Non-Biodegradable Microplastics in Soils: A Brief Review and Challenge." *Journal of Hazardous Materials* 409:124525. https://doi.org/10.1016/j.jhazmat.2020.124525.
84. Bergami, Elisa, Emilia Rota, Tancredi Caruso, Giovanni Birarda, Lisa

Vaccari, and Ilaria Corsi. 2020. "Plastics Everywhere: First Evidence of Polystyrene Fragments inside the Common Antarctic Collembolan *Cryptopygus antarcticus.*" *Biology Letters* 6 (6). https://doi.org/10.1098/rsbl.2020.0093.

85. Bourdages, Madelaine P. T., Jennifer F. Provencher, Julia E. Baak, Mark L. Mallory, and Jesse C. Vermaire. 2021. "Breeding Seabirds as Vectors of Microplastics from Sea to Land: Evidence from Colonies in Arctic Canada." *Science of the Total Environment* 764:142808. https://doi.org/10.1016/j.scitotenv.2020.142808.
86. Amélineau, Francoise, D. Bonnet, Oliver Heitz, Valentine Mortreux, Ann M. A. Harding, Nina Karnovsky, Wojciech Walkusz, et al. 2016. "Microplastic Pollution in the Greenland Sea: Background Levels and Selective Contamination of Planktivorous Diving Seabirds." *Environmental Pollution* 219:1131–39. http://dx.doi.org/10.1016/j.envpol.2016.09.017.
87. Le Guen, Camille, Giuseppe Suaria, Richard B. Sherley, Peter G. Ryan, Stefano Aliani, Lars Boehme, and Andrew S. Brierley. 2020. "Microplastic Study Reveals the Presence of Natural and Synthetic Fibres in the Diet of King Penguins (*Aptenodytes patagonicus*) Foraging from South Georgia." *Environment International* 134:105303. https://doi.org/10.1016/j.envint.2019.105303.
88. Audubon Guide to North American Birds. n.d. "Northern Fulmar." https://www.audubon.org/field-guide/bird/northern-fulmar; Audubon Guide to North American Birds. n.d. "Thick-billed Murre." https://www.audubon.org/field-guide/bird/thick-billed-murre.
89. Deng, Yanchun, Xuejian Jiang, Hongxia Zhao, Sa Yang, Jing Gao, Yanyan Wu, Qingyun Diao, and Chunsheng Hou. "Microplastic Polystyrene Ingestion Promotes the Susceptibility of Honeybee to Viral Infection." *Environmental Science and Technology* 55 (17): 11680–92. https://doi.org/10.1021/acs.est.1c01619.
90. Rader, R., Ignasi Bartomeus, Lucas A. Garibaldi, Michael P. D. Garratt, Brad G. Howlett, Rachael Winfree, Saul A. Cunningham, et al. 2016. "Non-Bee Insects Are Important Contributors to Global Crop Pollination." *Proceedings of the National Academy of Sciences* 113 (1): 146–51. https://doi.org/10.1073/pnas.1517092112.
91. Naggar, Al, Yahya, Markus Brinkmann, Christie M. Sayes, Saad N. AL-Kahtani, Showket A. Dar, Hesham R. El-Seedi, Bernd Grünewald, and John P. Giesy 2021. "Are Honey Bees at Risk From Microplastics?" *Toxics* 9 (5): 109. https://doi.org/10.3390/toxics9050109.
92. Al-Jaibachi, Rana, Ross N. Cuthbert, and Amanda Callaghan. 2018.

"Up and Away: Ontogenic Transference as a Pathway for Aerial Dispersal of Microplastics." *Royal Society Biology Letters* 14 (9). http://doi.org/10.1098/rsbl.2018.0479.

93. Boots, Bas, Connor William Russell, and Dannielle Senga Green. "Effects of Microplastics in Soil Ecosystems: Above and Below Ground." *Environmental Science and Technology* 53 (19), 11496–506. https://doi.org/10.1021/acs.est.9b03304.
94. Green, Dannielle S., Bas Boots, Jaime Da Silva Carvalho, and Thomas Starkey. 2019. "Cigarette Butts Have Adverse Effects on Initial Growth of Perennial Ryegrass (*Gramineae: Lolium perenne L.*) and White Clover (*Leguminosae: Trifolium repens L.*)." *Ecotoxicology and Environmental Safety* 182:109418. https://doi.org/10.1016/j.ecoenv.2019.109418.
95. McNear, David. 2013. "The Rhizosphere—Roots, Soil and Everything in Between." *Nature Education Knowledge* 4 (3): 1.
96. Sheldrake, Merlin. 2020. *Entangled Life: How Fungi Make Our Worlds, Change Our Minds, and Shape Our Futures.* New York: Penguin Random House.
97. Babikova, Zdenka, Lucy Gilbert, Toby J. A. Bruce, Michael Birkett, John C. Caulfield, Christine Woodcock, John A. Pickett, and David Johnson. 2013. "Underground Signals Carried through Common Mycelial Networks Warn Neighbouring Plants of Aphid Attack." *Ecology Letters* 16 (7): 835–43. https://doi.org/10.1111/ele.12115.
98. De Souza Machado, Anderson Abel, Chung W. Lau, Werner Kloas, Joana Bergmann, Julien B. Bachelier, Erik Faltin, Roland Becker, Anna S. Görlich, and Matthias C. Rillig. 2019. "Microplastics Can Change Soil Properties and Affect Plant Performance." *Environmental Science and Technology* 53 (10): 6044–52. https://doi.org/10.1021/acs.est.9b01339.
99. Lozano, Yudi M., Timon Lehnert, Lydia T. Linck, Anika Lehmann, and Matthias C. Rillig. 2021. "Microplastic Shape, Polymer Type, and Concentration Affect Soil Properties and Plant Biomass." *Frontiers in Plant Science* 12. https://doi.org/10.3389/fpls.2021.616645.
100. Qi, Yueling, Xiaomei Yang, Amalia Mejia Pelaez, Esperanza Huerta Lwanga, Nicolas Beriot, Henny Gertsen, Paolina Garbeva, and Violette Geissen. 2018. "Macro- and Micro- Plastics in Soil-Plant System: Effects of Plastic Mulch Film Residues on Wheat (*Triticum aestivum*) Growth." *Science of the Total Environment* 645:1048–56. https://doi.org/10.1016/j.scitotenv.2018.07.229.
101. Qi, Yueling, Adam Ossowicki, Xiaomei Yang, Esperanza Huerta Lwanga, Francisco Dini-Andreote, Violette Geissen, and Paolina

Garbeva. 2020. "Effects of Plastic Mulch Film Residues on Wheat Rhizosphere and Soil Properties." *Journal of Hazardous Materials* 387:121711. https://doi.org/10.1016/j.jhazmat.2019.121711; Sintim, Henry Y., Sreejata Bandopadhyay, Marie E. English, Andy Bary, José E. Liquet y González, Jennifer M. DeBruyn, Sean M. Schaeffer, Carol A. Miles, and Markus Flury. 2021. "Four Years of Continuous Use of Soil-Biodegradable Plastic Mulch: Impact on Soil and Groundwater Quality." *Geoderma* 381:114665. https://doi.org/10.1016/j.geoderma.2020.114665.

102. Taylor, Stephen E., Carolyn I. Pearce, Karen A. Sanguinet, Dehong Hu, William B. Chrisler, Young-Mo Kim, Zhan Wange, and Markus Flury. 2020. "Polystyrene Nano- and Microplastic Accumulation at Arabidopsis and Wheat Root Cap Cells, but No Evidence for Uptake into Roots." *Environmental Science: Nano* 7 (7): 1942–53. https://doi.org/10.1039/D0EN00309C.
103. Mateos-Cárdenas, Alicia, Frank N. A. M. van Pelt, John O'Halloran, and Marcel A. K. Jansen. 2021. "Adsorption, Uptake and Toxicity of Micro- and Nanoplastics: Effects on Terrestrial Plants and Aquatic Macrophytes." *Environmental Pollution* 284:117183. https://doi.org/10.1016/j.envpol.2021.117183.
104. Dasgupta, Shreya. 2016. "How Many Plant Species Are There in the World? Scientists Now Have an Answer." *Mongabay*. https://news.mongabay.com/2016/05/many-plants-world-scientists-may-now-answer/.
105. Bosker, Thijs, Lotte J. Bouwman, Nadja R. Brun, Paul Behrens, and Martina G. Vijver. "Microplastics Accumulate on Pores in Seed Capsule and Delay Germination and Root Growth of the Terrestrial Vascular Plant *Lepidium sativum*." *Chemosphere* 226:774–81. https://doi.org/10.1016/j.chemosphere.2019.03.163.
106. Helmberger, Maxwell S., Lisa K. Tiemann, and Matthew J. Grieshop. 2019. "Towards an Ecology of Soil Microplastics." *Functional Ecology* 34 (3). https://doi.org/10.1111/1365-2435.13495.
107. Lwanga, Esperanza Huerta, Jorge Mendoza Vega, Victor Ku Quej, Jesus de los Angeles Chi, Lucero Sanchez del Cid, Cesar Chi, Griselda Escalona Segura, et al. 2017. "Field Evidence for Transfer of Plastic Debris along a Terrestrial Food Chain." *Scientific Reports* 7:14071. https://www.nature.com/articles/s41598-017-14588-2.
108. Jacob, Jacquie. 2013. "Including Fishmeal in Organic Poultry Diets." eOrganic. https://eorganic.org/node/8129.
109. Diaz-Basantes, Milene F., Juan A. Conesa, and Andres Fullana. 2020.

"Microplastics in Honey, Beer, Milk and Refreshments in Ecuador as Emerging Contaminants." *Sustainability* 12 (14): 5514. https://doi.org/10.3390/su12145514.

110. Edo, Carlos, Amadeo R. Fernández-Alba, Flemming Vejsnæs, Jozef J. M. van der Steen, Francisca Fernández-Piñas, and Roberto Rosal. 2021. "Honeybees as Active Samplers for Microplastics." *Science of the Total Environment* 767:144481. https://doi.org/10.1016/j.scitotenv.2020.144481.
111. Kosuth, Mary, Sherri A. Mason, Elizabeth V. Wattenberg. 2018. "Anthropogenic Contamination of Tap Water, Beer, and Sea Salt." *PLoS ONE* 13 (4): e0194970. https://doi.org/10.1371/journal.pone.0194970.
112. Yang, Dongqi, Huahong Shi, Lan Li, Jiana Li, Khalida Jabeen, and Prabhu Kolandhasamy. 2015. "Microplastic Pollution in Table Salts from China." *Environmental Science and Technology* 49 (22): 13622–27. https://doi.org/10.1021/acs.est.5b03163.
113. Danopoulos, Evangelos, Lauren Jenner, Maureen Twiddy, and Jeanette M. Rotchell. 2020. "Microplastic Contamination of Salt Intended for Human Consumption: A Systematic Review and Meta-Analysis." *SN Applied Sciences* 2:1950. https://link.springer.com/article/10.1007/s42452-020-03749-0.
114. Liebezeit, Gerd, and Elisabeth Liebezeit. 2014. "Synthetic Particles as Contaminants in German Beers." *Food Additives and Contaminants: Part A* 31 (9): 1574–78. https://doi.org/10.1080/19440049.2014.945099.
115. Danopoulos, Evangelos, Maureen Twiddy, and Jeanette M. Rotchell. (2020) "Microplastic Contamination of Drinking Water: A Systematic Review." *PLoS ONE* 15 (7): e0236838. https://doi.org/10.1371/journal.pone.0236838.
116. Schymanski, Darena, Christophe Goldbeck, Hans-Ulrich Humpf, and Peter Fürst. 2018. "Analysis of Microplastics in Water by Micro-Raman Spectroscopy: Release of Plastic Particles from Different Packaging into Mineral Water." *Water Research* 129:154–62. https://doi.org/10.1016/j.watres.2017.11.011.
117. Simon, Matt. 2019. "You've Been Drinking Microplastics, but Don't Worry—Yet." *Wired*. https://www.wired.com/story/microplastic-who-study/.
118. Kosuth, Mary, Elizabeth V. Wattenberg, Sherri A. Mason, Christopher Tyree, and Dan Morrison. 2017. "Synthetic Polymer Contamination in Global Drinking Water." Orb Media. https://orbmedia.org/stories/Invisibles_final_report.

119. California Legislature. 2018. "SB-1422 California Safe Drinking Water Act: Microplastics." https://leginfo.legislature.ca.gov/faces/billTextClient.xhtml?bill_id=201720180SB1422.
120. California Ocean Protection Council. 2022. *Statewide Microplastics Strategy: Understanding and Addressing Impacts to Protect Coastal and Ocean Health*. Sacramento: Ocean Protection Council. https://www.opc.ca.gov/webmaster/ftp/pdf/agenda_items/20220223/Item_6_Exhibit_A_Statewide_Microplastics_Strategy.pdf.
121. World Health Organization. 2019. "1 in 3 People Globally Do Not Have Access to Safe Drinking Water—UNICEF, WHO." https://www.who.int/news/item/18-06-2019-1-in-3-people-globally-do-not-have-access-to-safe-drinking-water-unicef-who.
122. Tanentzap, Andrew J., Samuel Cottingham, Jérémy Fonvielle, Isobel Riley, Lucy M. Walker, Samuel G. Woodman, Danai Kontou, et al. 2021. "Microplastics and Anthropogenic Fibre Concentrations in Lakes Reflect Surrounding Land Use." *PLoS Biology* 19 (9): e3001389. https://doi.org/10.1371/journal.pbio.3001389.
123. Munno, Keenan, Paul A. Helm, Chelsea Rochman, Tara George, and Donald A. Jackson. 2021. "Microplastic Contamination in Great Lakes Fish." *Conservation Biology* 36 (1): e13794. https://doi.org/10.1111/cobi.13794.
124. Li, Daoji, Guyu Peng, and Lixin Zhu. 2019. "Progress and Prospects of Marine Microplastic Research in China." *Anthropocene Coasts* 2 (1): 330–39. https://doi.org/10.1139/anc-2018-0014.
125. Samandra, Subharthe, Julia M. Johnston, Julia E. Jaeger, Bob Symons, Shay Xie, Matthew Currell, Amanda V. Ellis, and Bradley O. Clarke. 2022. "Microplastic Contamination of an Unconfined Groundwater Aquifer in Victoria, Australia." *Science of the Total Environment* 802:149727. https://doi.org/10.1016/j.scitotenv.2021.149727.
126. Simon, Matt. 2020. "Babies May Be Drinking Millions of Microplastic Particles a Day." *Wired*. https://www.wired.com/story/babies-may-be-drinking-millions-of-microplastic-particles-a-day/.
127. Li, Dunzhu, Yunhong Shi, Luming Yang, Liwen Xiao, Daniel K. Kehoe, Yurii K. Gun'ko, John J. Boland, and Jing Jing Wang. 2020. "Microplastic Release from the Degradation of Polypropylene Feeding Bottles during Infant Formula Preparation." *Nature Food* 1:746–754 https://doi.org/10.1038/s43016-020-00171-y.
128. Su, Yu, Xi Hu, Hongjie Tang, Kun Lu, Huimin Li, Sijin Liu, and Baoshan Xing. 2021. "Steam Disinfection Releases Micro(nano)plastics

from Silicone-Rubber Baby Teats as Examined by Optical Photothermal Infrared Microspectroscopy." *Nature Nanotechnology*. https://doi.org/10.1038/s41565-021-00998-x.
129. Junjie Zhang, Lei Wang, Leonardo Trasande, and Kurunthachalam Kannan. 2021. "Occurrence of Polyethylene Terephthalate and Polycarbonate Microplastics in Infant and Adult Feces." *Environmental Science and Technology* 8 (11): 989–94. https://doi.org/10.1021/acs.estlett.1c00559; Simon, Matt. 2021. "Baby Poop Is Loaded with Microplastics." *Wired*. https://www.wired.com/story/baby-poop-is-loaded-with-microplastics/.
130. Ranjan, Ved Prakash, Anuja Joseph, and Sudha Goel. 2021. "Microplastics and Other Harmful Substances Released from Disposable Paper Cups into Hot Water." *Journal of Hazardous Materials* 404 (B): 124118. https://doi.org/10.1016/j.jhazmat.2020.124118.
131. Busse, Kristin, Ingo Ebner, Hans-Ulrich Humpf, Natalia Ivleva, Andrea Kaeppler, Barbara E. Oßmann, and Darena Schymanski. "Comment on 'Plastic Teabags Release Billions of Microparticles and Nanoparticles into Tea.'" *Environmental Science and Technology* 54 (21): 14134–35. https://doi.org/10.1021/acs.est.0c03182.
132. Shi, Yunhong, Dunzhu Li, Liwen Xiao, Daragh Mullarkey, Daniel K. Kehoe, Emmet D. Sheerin, Sebastian Barwich, et al. 2022. "Real-World Natural Passivation Phenomena Can Limit Microplastic Generation in Water." *Chemical Engineering Journal* 428:132466. https://doi.org/10.1016/j.cej.2021.132466.
133. Cox, Kieran D., Garth A. Covernton, Hailey L. Davies, John F. Dower, Francis Juanes, and Sarah E. Dudas. 2019. "Human Consumption of Microplastics." *Environmental Science and Technology* 53 (12): 7068–74. https://doi.org/10.1021/acs.est.9b01517.
134. Schwabl, Philipp, Sebastian Köppel, Philipp Königshofer, Theresa Bucsics, Michael Trauner, Thomas Reiberger, and Bettina Liebmann. 2019. "Detection of Various Microplastics in Human Stool: A Prospective Case Series." *Annals of Internal Medicine* 171:453–57. https://doi.org/10.7326/M19-0618.
135. Wright, Stephanie, and Ian Mudway. 2019. "The Ins and Outs of Microplastics." *Annals of Internal Medicine* 171:514–15. https://doi.org/10.7326/M19-2474.
136. Zhang, Na, Yi Bin Li, Hai Rong He, Jian Fen Zhang, and Guan Sheng Ma. 2021. "You Are What You Eat: Microplastics in the Feces of Young Men Living in Beijing." *Science of the Total Environment* 767:144345. https://doi.org/10.1016/j.scitotenv.2020.144345.

137. Catarino, Ana I., Valeria Macchia, William G, Sanderson, Richard C. Thompson, and Theodore B. Henry. 2018. "Low Levels of Microplastics (MP) in Wild Mussels Indicate That MP Ingestion by Humans Is Minimal Compared to Exposure via Household Fibres Fallout during a Meal." *Environmental Pollution* 237:675–84. https://doi.org/10.1016/j.envpol.2018.02.069.

**Chapter 4: Breathe Deep the Plastic Air**

1. Materić, Dušan, Elke Ludewig, Dominik Brunner, Thomas Röckmann, and Rupert Holzinger. 2021. "Nanoplastics Tansport to the Remote, High-Altitude Alps." *Environmental Pollution* 288:117697. https://doi.org/10.1016/j.envpol.2021.117697; Yang, Huirong, Yinglin He, Yumeng Yan, Muhammad Junaid, and Jun Wang. 2021. "Characteristics, Toxic Effects, and Analytical Methods of Microplastics in the Atmosphere." *Nanomaterials* 11 (10): 2747. https://doi.org/10.3390/nano11102747.
2. Materić, Dušan, Helle Astrid Kjær, Paul Vallelonga, Jean-Louis Tison, Thomas Röckmann, and Rupert Holzinger. 2022. "Nanoplastics Measurements in Northern and Southern Polar Ice." *Environmental Research* 208:112741. https://doi.org/10.1016/j.envres.2022.112741.
3. Gray, Ellen. 2017. "NASA Satellite Reveals How Much Saharan Dust Feeds Amazon's Plants." NASA. https://www.nasa.gov/content/goddard/nasa-satellite-reveals-how-much-saharan-dust-feeds-amazon-s-plants; Gouin, Todd. 2021. "Addressing the Importance of Microplastic Particles as Vectors for Long-Range Transport of Chemical Contaminants: Perspective in Relation to Prioritizing Research and Regulatory Actions." *Microplastics and Nanoplastics* 1:14. https://doi.org/10.1186/s43591-021-00016-w.
4. González-Pleiter, Miguel, Carlos Edo, Ángeles Aguilera, Daniel Viúdez-Moreiras, Gerardo Pulido-Reyes, Elena González-Toril, Susana Osuna, et al. 2021. "Occurrence and Transport of Microplastics Sampled within and above the Planetary Boundary Layer." *Science of the Total Environment* 761:143213. https://doi.org/10.1016/j.scitotenv.2020.143213.
5. Stefánsson, Hlynur, Mark Peternell, Matthias Konrad-Schmolke, Hrafnhildur Hannesdóttir, Einar J. Ásbjörnsson, and Erik Sturkell. 2021. "Microplastics in Glaciers: First Results from the Vatnajökull Ice Cap." *Sustainability* 13 (8): 4183. https://doi.org/10.3390/su13084183; Abbasi, Sajjad, and Andrew Turner. 2021. "Dry and Wet Deposition of

Microplastics in a Semi-Arid Region (Shiraz, Iran)." *Science of the Total Environment* 786:147358. https://doi.org/10.1016/j.scitotenv.2021.147358.

6. Simon, Matt. 2019. "A Shocking Find Shows Just How Far Wind Can Carry Microplastics." *Wired.* https://www.wired.com/story/wind-microplastics/; Allen, Steve, Deonie Allen, Vernon R. Phoenix, Gaël Le Roux, Pilar Durántez Jiménez, Anaëlle Simonneau, Stéphane Binet, and Didier Galop. 2019. "Atmospheric Transport and Deposition of Microplastics in a Remote Mountain Catchment." *Nature Geoscience* 12:339–44. https://doi.org/10.1038/s41561-019-0335-5.
7. Hamilton, Bonnie M., Madelaine P. T. Bourdages, Catherine Geoffroy, Jesse C. Vermaire, Mark L. Mallory, Chelsea M. Rochman, and Jennifer F. Provencher. 2021. "Microplastics around an Arctic Seabird Colony: Particle Community Composition Varies across Environmental Matrices." *Science of the Total Environment* 773:145536. https://doi.org/10.1016/j.scitotenv.2021.145536.
8. Purwiyanto, Anna Ida Sunaryo, Tri Prartono, Etty Riani, Yuli Naulita, Muhammad Reza Cordova, and Alan Frendy Koropitan. 2022. "The Deposition of Atmospheric Microplastics in Jakarta-Indonesia: The Coastal Urban Area." *Marine Pollution Bulletin* 174:113195. https://doi.org/10.1016/j.marpolbul.2021.113195.
9. Brahney, Janice, Natalie Mahowald, Marje Prank, Gavin Cornwell, Zbigniew Klimont, Hitoshi Matsui, and Kimberly Ann Prather. 2021. "Constraining the Atmospheric Limb of the Plastic Cycle." *Proceedings of the National Academy of Sciences* 118 (16): e2020719118. https://doi.org/10.1073/pnas.2020719118; Simon, Matt. 2021. "Plastic Is Falling from the Sky. But Where's It Coming From?" *Wired.* https://www.wired.com/story/plastic-is-falling-from-the-sky/.
10. Prata, Joana C., Joana L. Castro, Joao P. da Costa, Mario Cerqueira, Armando C. Duarte, and Teresa Rocha-Santos. 2020. "Airborne Microplastics." In *Handbook of Microplastics in the Environment* edited by Teresa Rocha-Santos, Mónica Costa, and Catherine Mouneyrac. N.p.: Springer, Cham. https://doi.org/10.1007/978-3-030-10618-8_37-1.
11. Sommer, Frank, Volker Dietze, Anja Baum, Jan Sauer, Stefan Gilge, Christoph Maschowski, and Reto Gieré. 2018. "Tire Abrasion as a Major Source of Microplastics in the Environment." *Aerosol and Air Quality Research* 18:2014–28. https://doi.org/10.4209/aaqr.2018.03.0099.
12. Evangeliou, Nikolaos, Henrik Grythe, Zbigniew Klimont, C. Heyes, S. Eckhardt, and Susana Lopez-Aparicio. 2020. "Atmospheric Transport

Is a Major Pathway of Microplastics to Remote Regions." *Nature Communications* 11:3381. https://doi.org/10.1038/s41467-020-17201-9; Simon, Matt. 2020. "Your Car Is Spewing Microplastics That Blow around the World." *Wired*. https://www.wired.com/story/your-car-is-spewing-microplastics/.

13. Simon, Matt. 2019. "Microplastics Are Blowing into the Pristine Arctic." *Wired*. https://www.wired.com/story/microplastics-are-blowing-into-the-pristine-arctic/; Bergmann, Melanie, Sophia Mützel, Sebastian Primpke, Mine B. Tekman, Jürg Trachsel, and Gunnar Gerdts. 2019. "White and Wonderful? Microplastics Prevail in Snow from the Alps to the Arctic." *Science Advances* 5 (8): eaax1157. https://doi.org/10.1126/sciadv.aax1157.
14. Xu, Guorui, Lei Yang, Li Xu, and Jie Yang. 2022. "Soil Microplastic Pollution under Different Land Uses in Tropics, Southwestern China." *Chemosphere* 289:133176. https://doi.org/10.1016/j.chemosphere.2021.133176.
15. Liang, Ting, Zhiyuan Lei, Md. Tariful Islam Fuad, Qi Wang, Shichun Sun, James Kar-Hei Fang, and Xiaoshou Liu. 2022. "Distribution and Potential Sources of Microplastics in Sediments in Remote Lakes of Tibet, China." *Science of the Total Environment* 806 (2): 150526. https://doi.org/10.1016/j.scitotenv.2021.150526.
16. Dris, Rachid, Johnny Gasperi, Mohamed Saad, Cécile Mirande, and Bruno Tassin. 2016. "Synthetic Fibers in Atmospheric Fallout: A Source of Microplastics in the Environment?" *Marine Pollution Bulletin* 104 (1–2): 290–93. https://doi.org/10.1016/j.marpolbul.2016.01.006; Dris, Rachid, Johnny Gasperi, Vincent Rocher, Mohamed Saad, Nicolas Renault, and Bruno Tassin. 2015. "Microplastic Contamination in an Urban Area: A Case Study in Greater Paris." *Environmental Chemistry* 12:592–99. https://doi.org/10.1071/EN14167; Dris, Rachid, Johnny Gasperi, and Bruno Tassin. 2018. "Sources and Fate of Microplastics in Urban Areas: A Focus on Paris Megacity." In *Freshwater Microplastics: Emerging Environmental Contaminants?* The Handbook of Environmental Chemistry 58 edited by Martin Wagner and Scott Lambert. https://doi.org/10.1007/978-3-319-61615-5_11; Klein, Malin, and Elke K. Fischer. 2019. "Microplastic Abundance in Atmospheric Deposition within the Metropolitan Area of Hamburg, Germany." *Science of the Total Environment* 685:96–103. https://doi.org/10.1016/j.scitotenv.2019.05.405; Cai, Liqi, Jundong Wang, Jinping Peng, Zhi Tan, Zhiwei Zhan, Xiangling Tan, and Qiuqiang Chen. 2017. "Characteristic

of Microplastics in the Atmospheric Fallout from Dongguan City, China: Preliminary Research and First Evidence." *Environmental Science and Pollution Research* 24:24928–24935. https://doi.org/10.1007/s11356-017-0116-x.

17. Wright, S. L., J. Ulke, A. Font, K. L. A. Chan, and F. J. Kelly. 2020. "Atmospheric Microplastic Deposition in an Urban Environment and an Evaluation of Transport." *Environment International* 136:105411. https://doi.org/10.1016/j.envint.2019.105411; Truong, Tran-Nguyen-Sang, Emilie Strady, Thuy-Chung Kieu-Le, Quoc-Viet Tran, Thi-Minh-Tam Le, and Quoc-Thinh Thuong. 2021. "Microplastic in Atmospheric Fallouts of a Developing Southeast Asian Megacity under Tropical Climate." *Chemosphere* 272:129874. https://doi.org/10.1016/j.chemosphere.2021.129874.
18. Fuller, Gary. 2021. "It's Not Just Oceans: Scientists Find Plastic Is Also Polluting the Air." *Guardian*. https://www.theguardian.com/environment/2021/feb/26/not-just-oceans-plastic-polluting-air-delhi-smog.
19. Dehghani, Sharareh, Farid Moore, and Razegheh Akhbarizadeh. 2017. "Microplastic Pollution in Deposited Urban Dust, Tehran Metropolis, Iran." *Environmental Science and Pollution Research* 24:20360–71. https://doi.org/10.1007/s11356-017-9674-1.
20. Abbasi, Sajjad, Behnam Keshavarzi, Farid Moore, Andrew Turner, Frank J. Kelly, Ana Oliete Dominguez, and Neemat Jaafarzadeh. 2019. "Distribution and Potential Health Impacts of Microplastics and Microrubbers in Air and Street Dusts from Asaluyeh County, Iran." *Environmental Pollution* 244:153–64. https://doi.org/10.1016/j.envpol.2018.10.039.
21. Syafei, Arie Dipareza, Nurul Rizki Nurasrin, Abdu Fadli Assomadi, and Rachmat Boedisantoso. 2019. "Microplastic Pollution in the Ambient Air of Surabaya, Indonesia." *Current World Environment* 14 (2). http://dx.doi.org/10.12944/CWE.14.2.13.
22. Allen, Steve, Deonie Allen, Kerry Moss, Gaël Le Roux, Vernon R. Phoenix, and Jeroen E. Sonke. 2020. "Examination of the Ocean as a Source for Atmospheric Microplastics." *PLoS ONE* 15 (5): e0232746. https://doi.org/10.1371/journal.pone.0232746.
23. Simon, Matt. 2020. "That Fresh Sea Breeze You Breathe May Be Laced with Microplastic." *Wired*. https://www.wired.com/story/sea-breeze-microplastic/.
24. Trainic, Miri, J. Michel Flores, Iddo Pinkas, Maria Luiza Pedrotti, Fabien Lombard, Guillaume Bourdin, Gabriel Gorsky, et al. 2020.

"Airborne Microplastic Particles Detected in the Remote Marine Atmosphere." *Communications Earth and Environment* 1:64. https://doi.org/10.1038/s43247-020-00061-y.

25. Lehmann, Moritz, Lisa Marie Oehlschlägel, Fabian P. Häusl, Andreas Held, and Stephan Gekle. 2021. "Ejection of Marine Microplastics by Raindrops: A Computational and Experimental Study." *Microplastics and Nanoplastics* 1:18. https://doi.org/10.1186/s43591-021-00018-8.
26. Meaza, Idoia, Jennifer H. Toyoda, and John Pierce Wise Sr. 2021. "Microplastics in Sea Turtles, Marine Mammals and Humans: A One Environmental Health Perspective." *Frontiers in Environmental Science* 8:298. https://doi.org/10.3389/fenvs.2020.575614.
27. Allen, S., D. Allen, F. Baladima, V. R. Phoenix, J. L. Thomas, G. Le Roux, and J. E. Sonke. 2021. "Evidence of Free Tropospheric and Long-Range Transport of Microplastic at Pic du Midi Observatory." *Nature Communications* 12:7242. https://doi.org/10.1038/s41467-021-27454-7.
28. Allen, D., S. Allen, G. Le Roux, A. Simonneau, D. Galop, and V. R. Phoenix. 2021. "Temporal Archive of Atmospheric Microplastic Deposition Presented in Ombrotrophic Peat." *Environmental Science and Technology Letters* 8. https://doi.org/10.1021/acs.estlett.1c00697.
29. Ganguly, Mainak, and Parisa A. Ariya. 2019. "Ice Nucleation of Model Nanoplastics and Microplastics: A Novel Synthetic Protocol and the Influence of Particle Capping at Diverse Atmospheric Environments." *ACS Earth and Space Chemistry* 3 (9): 1729–39. https://doi.org/10.1021/acsearthspacechem.9b00132; Zhang, Yulan, Shichang Kang, Steve Allen, Deonie Allen, Tanguang Gao, Mika Sillanpää. 2020. "Atmospheric Microplastics: A Review on Current Status and Perspectives." *Earth-Science Reviews* 203:103118. https://doi.org/10.1016/j.earscirev.2020.103118.
30. Revell, Laura E., Peter Kuma, Eric C. Le Ru, Walter R. C. Somerville, and Sally Gaw. 2021. "Direct Radiative Effects of Airborne Microplastics." *Nature* 598, 462–67. https://doi.org/10.1038/s41586-021-03864-x; Simon, Matt. 2021. "Microplastics May Be Cooling—and Heating—Earth's Climate." *Wired.* https://www.wired.com/story/microplastics-may-be-cooling-and-heating-earths-climate/.
31. Beaurepaire, Max, Rachid Dris, Johnny Gasperi, and Bruno Tassin. 2021. "Microplastics in the Atmospheric Compartment: A Comprehensive Review on Methods, Results on Their Occurrence and Determining Factors." *Current Opinion in Food Science.* https://doi.org/10.1016/j.cofs.2021.04.010; Dris, Rachid, Johnny Gasperi, Cécile Mirande,

Corinne Mandin, Mohamed Guerrouache, Valérie Langlois, and Bruno Tassin. 2017. "A First Overview of Textile Fibers, Including Microplastics, in Indoor and Outdoor Environments." *Environmental Pollution* 221:453–58. https://doi.org/10.1016/j.envpol.2016.12.013.

32. Liu, Chunguang, Jia Li, Yilei Zhang, Lei Wang, Jie Deng, Yuan Gao, Lu Yu, Junjie Zhang, and Hongwen Sun. 2019. "Widespread Distribution of PET and PC Microplastics in Dust in Urban China and Their Estimated Human Exposure." *Environment International* 128:116–24. https://doi.org/10.1016/j.envint.2019.04.024; Yao, Ying, Mihaela Glamoclija, Ashley Murphy, and Yuan Gao. 2021. "Characterization of Microplastics in Indoor and Ambient Air in Northern New Jersey. *Environmental Research* 207:112142. https://doi.org/10.1016/j.envres.2021.112142.
33. Gaston, Emily, Mary Woo, Clare Steele, Suja Sukumaran, and Sean Anderson. 2020. "Microplastics Differ between Indoor and Outdoor Air Masses: Insights from Multiple Microscopy Methodologies." *Applied Spectroscopy* 74 (9): 1079–98. https://doi.org/10.1177%2F0003702820920652.
34. Cai, Yaping, Denise M. Mitrano, Rudolf Hufenus, and Bernd Nowack. 2021. "Formation of Fiber Fragments during Abrasion of Polyester Textiles." *Environmental Science and Technology* 55 (12): 8001–8009. https://doi.org/10.1021/acs.est.1c00650.
35. De Falco, Francesca, Mariacristina Cocca, Maurizio Avella, and Richard C. Thompson. 2020. "Microfiber Release to Water, Via Laundering, and to Air, via Everyday Use: A Comparison between Polyester Clothing with Differing Textile Parameters." *Environmental Science and Technology* 54 (6). https://doi.org/10.1021/acs.est.9b06892.
36. Qun Zhang, Yaping Zhao, Fangni Du, Huiwen Cai, Gehui Wang, and Huahong Shi. 2020. "Microplastic Fallout in Different Indoor Environments." *Environmental Science and Technology* 54 (11): 6530–39. https://doi.org/10.1021/acs.est.0c00087.
37. Chen, Yingxin, Xinyu Li, Xiaoting Zhang, Yalin Zhang, Wei Gao, Ruibin Wang, and Defu He. 2022. "Air Conditioner Filters Become Sinks and Sources of Indoor Microplastics Fibers." *Environmental Pollution* 292 (B): 118465. https://doi.org/10.1016/j.envpol.2021.118465.
38. Soltani, Neda Sharifi, Mark Patrick Taylor, and Scott Paton Wilson. 2021. "Quantification and Exposure Assessment of Microplastics in Australian Indoor House Dust." *Environmental Pollution* 283:117064. https://doi.org/10.1016/j.envpol.2021.117064.

39. Ageel, Hassan Khalid, Stuart Harrad, and Mohamed Abou-Elwafa Abdallah. 2021. "Occurrence, Human Exposure, and Risk of Microplastics in the Indoor Environment." *Environmental Science: Processes and Impacts*. https://doi.org/10.1039/D1EM00301A.
40. Vianello, Alvise, Rasmus Lund Jensen, Li Liu, and Jes Vollertsen. 2019. "Simulating Human Exposure to Indoor Airborne Microplastics Using a Breathing Thermal Manikin." *Scientific Reports* 9:8670. https://doi.org/10.1038/s41598-019-45054-w.
41. Winkens, Kerstin, Robin Vestergren, Urs Berger, and Ian T. Cousins. 2017. "Early Life Exposure to Per- and Polyfluoroalkyl Substances (PFASs): A Critical Review." *Emerging Contaminants* 3 (2): 55–68. https://doi.org/10.1016/j.emcon.2017.05.001.
42. Torres-Agullo, A., A. Karanasiou, T. Moreno, and S. Lacorte. 2021. "Overview on the Occurrence of Microplastics in Air and Implications from the Use of Face Masks During the COVID-19 Pandemic." *Science of the Total Environment* 800:149555. https://doi.org/10.1016/j.scitotenv.2021.149555.
43. Li, Lu, Xiaoli Zhao, Zhouyang Li, and Kang Song. 2021. "COVID-19: Performance Study of Microplastic Inhalation Risk Posed by Wearing Masks." *Journal of Hazardous Materials* 411:124955. https://doi.org/10.1016/j.jhazmat.2020.124955.
44. OceansAsia. 2020. "COVID-19 Facemasks and Marine Plastic Pollution." https://oceansasia.org/covid-19-facemasks/; Wang, Zheng, Chunjiang An, Xiujuan Chen, Kenneth Lee, Baiyu Zhang, and Qi Feng. 2021. "Disposable Masks Release Microplastics to the Aqueous Environment with Exacerbation by Natural Weathering." *Journal of Hazardous Materials* 417:126036. https://doi.org/10.1016/j.jhazmat.2021.126036.
45. Sobhani, Zahra, Yongjia Lei, Youhong Tang, Liwei Wu, Xian Zhang, Ravi Naidu, Mallavarapu Megharaj, and Cheng Fang. 2020. "Microplastics Generated When Opening Plastic Packaging." *Scientific Reports* 10:4841. https://doi.org/10.1038/s41598-020-61146-4.
46. O'Brien, Stacey, Elvis D. Okoffo, Jake W. O'Brien, Francisca Ribeiro, Xianyu Wang, Stephanie L. Wright, Saer Samanipour, et al. 2020. "Airborne Emissions of Microplastic Fibres from Domestic Laundry Dryers." *Science of the Total Environment* 747:141175. https://doi.org/10.1016/j.scitotenv.2020.141175.
47. Kapp, Kirsten J., and Rachael Z. Miller. 2020. "Electric Clothes Dryers: An Underestimated Source of Microfiber Pollution." *PLoS ONE*. 15 (10): e0239165. https://doi.org/10.1371/journal.pone.0239165.

48. Tao, Danyang, Kai Zhang, Shaopeng Xu, Huiju Lin, Yuan Liu, Jingliang Kang, Tszewai Yim, et al. 2022. "Microfibers Released into the Air from a Household Tumble Dryer." *Environmental Science and Technology Letters.* http://dx.doi.org/10.1021/acs.estlett.1c00911.
49. Nor, Nur Hazimah Mohamed, Merel Kooi, Noël J. Diepens, and Albert A. Koelmans. 2021. "Lifetime Accumulation of Microplastic in Children and Adults." *Environmental Science and Technology* 55 (8): 5084–96. https://doi.org/10.1021/acs.est.0c07384.
50. The Cleveland Clinic. n.d. "Lungs: How They Work." https://my.clevelandclinic.org/health/articles/8960-lungs-how-they-work.
51. Wieland, Simon, Aylin Balmes, Julian Bender, Jonas Kitzinger, Felix Meyer, Anja F. R. M. Ramsperger, Franz Roeder, Caroline Tengelmann, Benedikt H. Wimmer, Christian Laforsch, and Holger Kress. 2022. "From Properties to Toxicity: Comparing Microplastics to Other Airborne Microparticles." *Journal of Hazardous Materials* 428:128151. https://doi.org/10.1016/j.jhazmat.2021.128151.
52. Pauly, John L., Sharon J. Stegmeier, Heather A. Allaart, Richard T. Cheney, Paul J. Zhang, Andrew G. Mayer, and Richard J. Streck. 1998. "Inhaled Cellulosic and Plastic Fibers Found in Human Lung Tissue." *Cancer Epidemiology, Biomarkers, and Prevention* 7:419e428.
53. Prata, Joana Correia. 2018. "Airborne Microplastics: Consequences to Human Health?" *Environmental Pollution* 234:115–26. https://doi.org/10.1016/j.envpol.2017.11.043.
54. Zarus, Gregory M., Custodio Muianga, Candis M. Hunter, and R. Steven Pappas. 2021. "A Review of Data for Quantifying Human Exposures to Micro and Nanoplastics and Potential Health Risks." *Science of the Total Environment* 756:144010. https://doi.org/10.1016/j.scitotenv.2020.144010.
55. Pimentel, J. Cortez, Ramiro Avila, and A. Galvao Lourenço. 1975. "Respiratory Disease Caused by Synthetic Fibres: A New Occupational Disease." *Thorax* 30 (2): 204–19. https://doi.org/10.1136/thx.30.2.204.
56. Mastrangelo, Giuseppe, Ugo Fedeli, Emanuela Fadda, Giovanni Milan, and John H. Lange. 2020. "Epidemiologic Evidence of Cancer Risk in Textile Industry Workers: A Review and Update." *Toxicology and Industrial Health* 18 (4). https://doi.org/10.1191%2F0748233702th139rr; Vobecky, Josef, Ghislain Devroede, Jacques Lacaille, and Alain Watier. 1978. "An Occupational Group with a High Risk of Large Bowel Cancer." *Gastroenterology* 75 (2): 221–23. https://doi.org/10.1016/0016-5085(78)90406-7.

57. Swicofil AG. n.d. "Flock and the Flocking Process." https://www.swicofil.com/consult/industrial-applications/para-textil-and-carpets/flock; Facciolà, Alessio, Giuseppa Visalli, Marianna Pruiti Ciarello, and Angela Di Pietro. 2021. "Newly Emerging Airborne Pollutants: Current Knowledge of Health Impact of Micro and Nanoplastics." *International Journal of Environmental Research and Public Health* 18 (6): 2997. https://doi.org/10.3390/ijerph18062997.
58. Kern, David G., Robert S. Crausman, and Kate T. H. Durand. 1998. "Flock Worker's Lung: Chronic Interstitial Lung Disease in the Nylon Flocking Industry." *Annals of Internal Medicine* 129 (4): 261–72. https://doi.org/10.7326/0003-4819-129-4-199808150-00001.
59. Brandt-Rauf, Paul Wesley, Yongliang Li, Changmin Long, Regina Monaco, Gopala Kovvali, and Marie-Jeanne Marion. 2012. "Plastics and Carcinogenesis: The Example of Vinyl Chloride." *Journal of Carcinogenesis* 11:5. https://doi.org/10.4103/1477-3163.93700; National Cancer Institute. 2018. "Vinyl Chloride." https://www.cancer.gov/about-cancer/causes-prevention/risk/substances/vinyl-chloride.
60. EPA (Environmental Protection Agency). n.d. "Acrylamide." https://www.epa.gov/sites/production/files/2016-09/documents/acrylamide.pdf; EPA. n.d. "Acrylonitrile." https://www.epa.gov/sites/production/files/2016-09/documents/acrylonitrile.pdf; EPA. n.d. "Epichlorohydrin (1-Chloro-2,3-Epoxypropane)." https://www.epa.gov/sites/production/files/2016-09/documents/epichlorohydrin.pdf; Galloway, Tamara S. 2015."Micro- and Nano-Plastics and Human Health." In *Marine Anthropogenic Litter* edited by Melanie Bergmann, Lars Gutow, and Michael Klages. N.p.: Springer. https://doi.org/10.1007/978-3-319-16510-3_13.
61. Campanale, Claudia, Carmine Massarelli, Ilaria Savino, Vito Locaputo, and Vito Felice Uricchio. 2020. "A Detailed Review Study on Potential Effects of Microplastics and Additives of Concern on Human Health." *International Journal of Environmental Research and Public Health* 17 (4): 1212. https://doi.org/10.3390/ijerph17041212.
62. EPA. 2021. "Plastic Pollution." https://www.epa.gov/trash-free-waters/plastic-pollution.
63. Eriksen, Marcus, Martin Thiel, Matt Prindiville, and Tim Kiessling. 2018. "Microplastic: What Are the Solutions?" In *Freshwater Microplastics: Emerging Environmental Contaminants?* The Handbook of Environmental Chemistry 58 edited by Martin Wagner and Scott Lambert. https://doi.org/10.1007/978-3-319-61615-5_11; Liu, Sitong,

Jiafu Shi, Jiao Wang, Yexin Dai, Hongyu Li, Jiayao Li, Xianhua Liu, Xiaochen Chen, Zhiyun Wang, and Pingping Zhang. 2021. "Interactions between Microplastics and Heavy Metals in Aquatic Environments: A Review." *Frontiers in Microbiology* 12:652520. https://doi.org/10.3389/fmicb.2021.652520; Naqash, Nafiaah, Sadguru Prakash, Dhriti Kapoor, and Rahul Singh. 2020. "Interaction of Freshwater Microplastics with Biota and Heavy Metals: A Review." *Environmental Chemistry Letters* 18:1813–24. https://doi.org/10.1007/s10311-020-01044-3.

64. Mato, Yukie, Tomohiko Isobe, Hideshige Takada, Haruyuki Kanehiro, Chiyoko Ohtake, and Tsuguchika Kaminuma. 2001. "Plastic Resin Pellets as a Transport Medium for Toxic Chemicals in the Marine Environment." *Environmental Science and Technology* 35 (2): 318–24. https://doi.org/10.1021/es0010498.
65. Bakir, Adil, Steven J. Rowland, and Richard C. Thompson. 2014. "Transport of Persistent Organic Pollutants by Microplastics in Estuarine Conditions." *Estuarine, Coastal, and Shelf Science* 140:14–21. https://doi.org/10.1016/j.ecss.2014.01.004; EPA. 1972. "DDT Ban Takes Effect." https://archive.epa.gov/epa/aboutepa/ddt-ban-takes-effect.html.
66. Amato-Lourenço, Luís Fernando, Regiani Carvalho-Oliveira, Gabriel Ribeiro Júnior, Luciana dos Santos Galvão, Rômulo Augusto Ando, and Thais Mauad. 2021. "Presence of Airborne Microplastics in Human Lung Tissue." *Journal of Hazardous Materials* 416:126124. https://doi.org/10.1016/j.jhazmat.2021.126124.
67. Jenner, Lauren C., Jeanette M. Rotchell, Robert T. Bennett, Michael Cowen, Vasileios Tentzeris, and Laura R. Sadofsky. 2022. "Detection of Microplastics in Human Lung Tissue Using μFTIR Spectroscopy." *Science of the Total Environment* 831:154907. https://doi.org/10.1016/j.scitotenv.2022.154907.
68. Van Dijk, Fransien, Shan Shan Song, Gail van Eck, Xin Hui Wu, I. S. T. Bos, Devin Boom, Ingeborg M. Kooter, et al. 2021. "Inhalable Textile Microplastic Fibers Impair Airway Epithelial Growth." *bioRxiv*. https://doi.org/10.1101/2021.01.25.428144.
69. Centers for Disease Control and Prevention. 2011. "Asthma in the US." https://www.cdc.gov/vitalsigns/asthma/index.html.
70. Ritchie, Hannah, and Max Roser. 2018. "Plastic Pollution." Our World in Data. https://ourworldindata.org/plastic-pollution.
71. Lu, Kuo, Keng Po Lai, Tobias Stoeger, Shuqin Ji, Ziyi Lin, Xiao Lin, Ting Fung Chan, et al. 2021. "Detrimental Effects of Microplastic

Exposure on Normal and Asthmatic Pulmonary Physiology." *Journal of Hazardous Materials* 416:126069. https://doi.org/10.1016/j.jhazmat.2021.126069.

72. Wood, Trina. 2021. "Air Quality Linked to Increased Risk of Alzheimer's." University of California, Davis. https://www.vetmed.ucdavis.edu/news/air-quality-linked-increased-risk-alzheimers; Chen, Hong, Jeffrey C. Kwong, Ray Copes, Karen Tu, Paul J. Villeneuve, Aaron van Donkelaar, Perry Hystad, et al. 2017. "Living Near Major Roads and the Incidence of Dementia, Parkinson's Disease, and Multiple Sclerosis: A Population-Based Cohort Study." *Lancet* 389 (10070): 718–26. https://doi.org/10.1016/S0140-6736(16)32399-6.
73. Prata, Joana Correia Prata, João P. da Costa, Isabel Lopes, Armando C. Duarte, and Teresa Rocha-Santos. 2020. "Environmental Exposure to Microplastics: An Overview on Possible Human Health Effects." *Science of the Total Environment* 702:134455. https://doi.org/10.1016/j.scitotenv.2019.134455.
74. Sripada, Kam, Aneta Wierzbicka, Khaled Abass, Joan O. Grimalt, Andreas Erbe, Halina B. Röllin, Pál Weihe, et al. 2022. "A Children's Health Perspective on Nano- and Microplastics." *Environmental Health Perspectives* 130 (1). https://doi.org/10.1289/EHP9086.
75. Flaws, Jodi, Pauliina Damdimopoulou, Heather B. Patisaul, Andrea Gore, Lori Raetzman, and Laura N. Vandenberg. 2020. *Plastics, EDCs, and Health: Authoritative Guide*. Washington, DC: Endocrine Society. https://www.endocrine.org/topics/edc/plastics-edcs-and-health.
76. Environmental Working Group. n.d. "BPA." https://www.ewg.org/areas-focus/toxic-chemicals/bpa.
77. Vandenberg, Laura N. 2011. "Exposure to Bisphenol A in Canada: Invoking the Precautionary Principle." *Canadian Medical Association Journal* 183 (11): 1265–70. https://doi.org/10.1503/cmaj.101408.
78. FDA (Food and Drug Administration). n.d. "Bisphenol A (BPA): Use in Food Contact Application." https://www.fda.gov/food/food-additives-petitions/bisphenol-bpa-use-food-contact-application.
79. Asimakopoulos, Alexandros G., Madhavan Elangovan, and Kurunthachalam Kannan. 2016. "Migration of Parabens, Bisphenols, Benzophenone-Type UV Filters, Triclosan, and Triclocarban from Teethers and Its Implications for Infant Exposure." *Environmental Science and Technology* 50 (24): 13539–47. https://doi.org/10.1021/acs.est.6b04128.
80. Wang, Lei, Yilei Zhang, Yubin Liu, Xinying Gong, Tao Zhang, and Hongwen Sun. 2019. "Widespread Occurrence of Bisphenol A in Daily

Clothes and Its High Exposure Risk in Humans." *Environmental Science and Technology* 53 (12): 7095–102. https://doi.org/10.1021/acs.est.9b02090; Xue, Jingchuan, Wenbin Liu, and Kurunthachalam Kannan. 2017. "Bisphenols, Benzophenones, and Bisphenol A Diglycidyl Ethers in Textiles and Infant Clothing." *Environmental Science and Technology* 51 (9): 5279–86. https://doi.org/10.1021/acs.est.7b00701; Sait, Shannen T. L, Lisbet Sørensen, Stephan Kubowicz, Kristine Vike-Jonas, Susana V. Gonzalez, Alexandros G. Asimakopoulos, and Andy M. Booth. 2021. "Microplastic Fibres from Synthetic Textiles: Environmental Degradation and Additive Chemical Content." *Environmental Pollution* 268 (B): 115745. https://doi.org/10.1016/j.envpol.2020.115745.

81. DiFrisco, Emily. 2021. "CEH Finds 63 Sock Brands with High Levels of BPA." Center for Environmental Health. https://ceh.org/ceh-finds-63-sock-brands-with-high-levels-of-bpa/.
82. Vogel, Sarah. 2009. "The Politics of Plastics: The Making and Unmaking of Bisphenol A 'Safety.'" *American Journal of Public Health* 99 (S3): S559–S566. https://dx.doi.org/10.2105%2FAJPH.2008.159228.
83. Giuliani, Angela, Mariachiara Zuccarini, Angelo Cichelli, Haroon Khan, and Marcella Reale. 2020. "Critical Review on the Presence of Phthalates in Food and Evidence of Their Biological Impact." *International Journal of Environmental Research and Public Health* 17 (16): 5655. https://doi.org/10.3390/ijerph17165655.
84. Net, Sopheak, Richard Sempéré, Anne Delmont, Andrea Paluselli, and Baghdad Ouddane. 2015. "Occurrence, Fate and Behavior and Ecotoxicological State of Phthalates in Different Environmental Matrices." *Environmental Science and Technology* 49 (7): 4019–35. https://doi.org/10.1021/es505233b.
85. Hlisníková, Henrieta, Ida Petrovičová, Branislav Kolena, Miroslava Šidlovská, and Alexander Sirotkin. 2020. "Effects and Mechanisms of Phthalates' Action on Reproductive Processes and Reproductive Health: A Literature Review." *International Journal of Environmental Research and Public Health* 17 (18): 6811. https://doi.org/10.3390/ijerph17186811.
86. Jacobson, Melanie H., Cheryl R. Stein, Mengling Liu, Marra G. Ackerman, Jennifer K. Blakemore, Sara E. Long, Graziano Pinna, et al. 2021. "Prenatal Exposure to Bisphenols and Phthalates and Postpartum Depression: The Role of Neurosteroid Hormone Disruption." *Journal of Clinical Endocrinology and Metabolism* 106 (7): 1887–99. https://doi.org/10.1210/clinem/dgab199.

87. Kurunthachalam, Kannan, and Vimalkumar Krishnamoorthi. 2021. "A Review of Human Exposure to Microplastics and Insights into Microplastics as Obesogens." *Frontiers in Endocrinology* 12. https://doi.org/10.3389/fendo.2021.724989.
88. Völker, Johannes, Felicity Ashcroft, Åsa Vedøy, Lisa Zimmermann, and Martin Wagner. 2022. "Adipogenic Activity of Chemicals Used in Plastic Consumer Products." *Environmental Science and Technology.* https://doi.org/10.1021/acs.est.1c06316.
89. Trasande, Leonardo, Buyun Liu, and Wei Bao. 2021. "Phthalates and Attributable Mortality: A Population-Based Longitudinal Cohort Study and Cost Analysis." *Environmental Pollution* 118021. https://doi.org/10.1016/j.envpol.2021.118021.
90. Schreder, Erika, and Matthew Goldberg. 2022. "Toxic Convenience: The Hidden Costs of Forever Chemicals in Stain- and Water-Resistant Products." Toxic-Free Future. https://toxicfreefuture.org/pfas-in-stain-water-resistant-products-study/.
91. Herkert, Nicholas J., Christopher D. Kassotis, Sharon Zhang, Yuling Han, Vivek Francis Pulikkal, Mei Sun, P. Lee Ferguson, and Heather M. Stapleton. 2022. "Characterization of Per- and Polyfluorinated Alkyl Substances Present in Commercial Anti-Fog Products and Their In Vitro Adipogenic Activity." *Environmental Science and Technology*. https://doi.org/10.1021/acs.est.1c06990.
92. Bhagwat, Geetika, Thi Kim Anh Tran, Dane Lamb, Kala Senathirajah, Ian Grainge, Wayne O'Connor, Albert Juhasz, and Thava Palanisami. 2021. "Biofilms Enhance the Adsorption of Toxic Contaminants on Plastic Microfibers under Environmentally Relevant Conditions." *Environmental Science and Technology* 55 (13): 8877–87. https://doi.org/10.1021/acs.est.1c02012.
93. Burki, Talha. 2021. "PFAS: Here Today—Here Tomorrow." *Lancet* 9 (12). https://doi.org/10.1016/S2213-8587(21)00294-1.
94. Schirmer, Elisabeth, Stefan Schuster, and Peter Machnik. 2021. "Bisphenols Exert Detrimental Effects on Neuronal Signaling in Mature Vertebrate Brains." *Communications Biology*. https://doi.org/10.1038/s42003-021-01966-w.
95. National Cancer Institute. n.d. "Diethylstilbestrol (DES) and Cancer." https://www.cancer.gov/about-cancer/causes-prevention/risk/hormones/des-fact-sheet.
96. Ragusa, Antonio, Alessandro Svelato, Criselda Santacroce, Piera Catalano, Valentina Notarstefano, Oliana Carnevali, Fabrizio Papa, et al.

2021. "Plasticenta: First Evidence of Microplastics in Human Placenta." *Environment International* 146:106274. https://doi.org/10.1016/j.envint.2020.106274.

97. Braun, Thorsten, Loreen Ehrlich, Wolfgang Henrich, Sebastian Koeppel, Ievgeniia Lomako, Philipp Schwabl, and Bettina Liebmann. 2021. "Detection of Microplastic in Human Placenta and Meconium in a Clinical Setting" *Pharmaceutics* 13 (7): 921. https://doi.org/10.3390/pharmaceutics13070921.
98. Briffa, Sophie M. 2021. "Looking at the Bigger Picture—Considering the Hurdles in the Struggle against Nanoplastic Pollution." *Nanomaterials* 11 (10): 2536. https://doi.org/10.3390/nano11102536; Fournier, Sara B., Jeanine N. D'Errico, Derek S. Adler, Stamatina Kollontzi, Michael J. Goedken, Laura Fabris, Edward J. Yurkow, and Phoebe A. Stapleton. 2020. "Nanopolystyrene Translocation and Fetal Deposition after Acute Lung Exposure during Late-Stage Pregnancy." *Particle and Fibre Toxicology* 17:55. https://doi.org/10.1186/s12989-020-00385-9.
99. Wiesinger, Helene, Zhanyun Wang, and Stefanie Hellweg. 2021. "Deep Dive into Plastic Monomers, Additives, and Processing Aids." *Environmental Science and Technology* 55 (13): 9339–51. https://doi.org/10.1021/acs.est.1c00976.
100. Ibrahim, Yusof Shuaib, Sabiqah Tuan Anuar, Alyza A. Azmi, Wan Mohd Afiq Wan Mohd Khalik, Shumpei Lehata, Siti Rabaah Hamzah, Dzulkiflee Ismail, et al. 2021. "Detection of Microplastics in Human Colectomy Specimens." *JGH Open* 5:116–21. https://doi.org/10.1002/jgh3.12457.
101. Hildebrandt, Lars, Helmholtz-Zentrum Hereon, Fenna Nack, Tristan Zimmermann, Helmholtz-Zentrum Hereon, and Daniel Pröfrock. 2021. "Microplastics as a Trojan Horse for Trace Metals." *Journal of Hazardous Materials Letters* 2:100035. https://doi.org/10.1016/j.hazl.2021.100035.
102. Leslie, Heather A., Martin J. M. van Velzen, Sicco H. Brandsma, Dick Vethaak, Juan J. Garcia-Vallejo, and Marja H. Lamoree. 2022. "Discovery and Quantification of Plastic Particle Pollution in Human Blood." *Environment International* 107199. https://doi.org/10.1016/j.envint.2022.107199.
103. Lett, Zachary, Abigail Hall, Shelby Skidmore, and Nathan J. Alves. 2021. "Environmental Microplastic and Nanoplastic: Exposure Routes and Effects on Coagulation and the Cardiovascular System. *Environmental Pollution* 291:118190. https://doi.org/10.1016/j.envpol.2021.118190.

104. Gkoutselis, Gerasimos, Stephan Rohrbach, Janno Harjes, Martin Obst, Andreas Brachmann, and Marcus A. Horn. 2021. "Microplastics Accumulate Fungal Pathogens in Terrestrial Ecosystems." *Scientific Reports* 11:13214. https://doi.org/10.1038/s41598-021-92405-7.
105. Aghaei Gharehbolagh, S., M. Nasimi, S. Agha Kuchak Afshari, Z. Ghasemi, and S. Rezaie. 2017. "First Case of Superficial Infection Due to *Naganishia albida* (Formerly *Cryptococcus albidus*) in Iran: A Review of the Literature." *Current Medical Mycology* 3 (2): 33–37. https://doi.org/10.18869/acadpub.cmm.3.2.33.
106. Tamargo, Alba, Natalia Molinero, Julián J. Reinosa, Victor Alcolea-Rodriguez, Raquel Portela, Miguel A. Bañares, Jose F. Fernández, and M. Victoria Moreno-Arribas. 2022. "PET Microplastics Affect Human Gut Microbiota Communities during Simulated Gastrointestinal Digestion, First Evidence of Plausible Polymer Biodegradation during Human Digestion." *Scientific Reports* 12:528. https://doi.org/10.1038/s41598-021-04489-w.
107. Mayo Clinic. "Inflammatory Bowel Disease." https://www.mayoclinic.org/diseases-conditions/inflammatory-bowel-disease/symptoms-causes/syc-20353315; Yan, Zehua, Yafei Liu, Ting Zhang, Faming Zhang, Hongqiang Ren, and Yan Zhang. 2021. "Analysis of Microplastics in Human Feces Reveals a Correlation between Fecal Microplastics and Inflammatory Bowel Disease Status." *Environmental Science and Technology.* https://doi.org/10.1021/acs.est.1c03924.
108. Carrington, Damian. 2021. "Microplastics May Be Linked to Inflammatory Bowel Disease, Study Finds." *Guardian.* https://www.theguardian.com/society/2021/dec/22/microplastics-may-be-linked-to-inflammatory-bowel-disease-study-finds.
109. Van Megchelen, Pieter, and Dick Vethaak. 2020. *What Are Microplastics Doing in Our Bodies? A Knowledge Agenda for Microplastics and Health.* The Hague: Netherlands Organisation for Health Research and Development.
110. Urban, Robert M., Joshua Jacobs, Michael Tomlinson, John Gavrilovic, Jonathan Black, and Michel Peoc'h. 2000. "Dissemination of Wear Particles to the Liver, Spleen, and Abdominal Lymph Nodes of Patients with Hip or Knee Replacement." *Journal of Bone and Joint Surgery* 82 (4): 457. https://doi.org/10.2106/00004623-200004000-00002.
111. Prüst, Minne, Jonelle Meijer, and Remco H. S. Westerink. 2020. "The Plastic Brain: Neurotoxicity of Micro- and Nanoplastics." *Particle and Fibre Toxicology* 17:24. https://doi.org/10.1186/s12989-020-00358-y.
112. Kwon, Wookbong, Daehwan Kim, Hee-Yeon Kim, Sang Won Jeong,

Se-Guen Lee, Hyun-Chul Kim, Young-Jae Lee, et al. 2022. "Microglial Phagocytosis of Polystyrene Microplastics Results in Immune Alteration and Apoptosis in Vitro and in Vivo." *Science of the Total Environment* 807 (2): 150817. https://doi.org/10.1016/j.scitotenv.2021.150817.

113. University of York. 2021. "Microplastics Found to Be Harmful to Human Cells." https://www.york.ac.uk/news-and-events/news/2021/research/microplastics-harmful-human-cells/.
114. Science Advice for Policy by European Academies. 2019. "A Scientific Perspective on Microplastics in Nature and Society." https://doi.org/10.26356/microplastics.

**Chapter 5: Turning Down the Plastic Tap**

1. Simon, Matt. 2018. "A 600-Meter-Long Plastic Catcher Heads to Sea, but Scientists Are Skeptical." *Wired*. https://www.wired.com/story/ocean-cleanup-skeptical-scientists/.
2. Simon, Matt. 2019. "Ocean Cleanup's Plastic Catcher Is Busted. So What Now?" *Wired*. https://www.wired.com/story/ocean-cleanups-plastic-catcher/.
3. Shiffman, David. 2018. "I Asked 15 Ocean Plastic Pollution Experts about the Ocean Cleanup Project, and They Have Concerns." *Southern Fried Science*. http://www.southernfriedscience.com/i-asked-15-ocean-plastic-pollution-experts-about-the-ocean-cleanup-project-and-they-have-concerns/.
4. Kaufman, Matt. 2021. "The Carbon Footprint Sham." *Mashable*. https://mashable.com/feature/carbon-footprint-pr-campaign-sham; Supran, Geoffrey, and Naomi Oreskes. 2021. "The Forgotten Oil Ads That Told Us Climate Change Was Nothing." *Guardian*. https://www.theguardian.com/environment/2021/nov/18/the-forgotten-oil-ads-that-told-us-climate-change-was-nothing.
5. Sullivan, Laura. 2020. "How Big Oil Misled the Public into Believing Plastic Would Be Recycled." NPR. https://www.npr.org/2020/09/11/897692090/how-big-oil-misled-the-public-into-believing-plastic-would-be-recycled.
6. Chen, Junliang, Jing Wu, Peter C. Sherrell, Jun Chen, Huaping Wang, Wei-xian Zhang, and Jianping Yang. 2022. "How to Build a Microplastics-Free Environment: Strategies for Microplastics Degradation and Plastics Recycling." *Advanced Science* 9 (6). https://doi.org/10.1002/advs.202103764; Padervand, Mohsen, Eric Lichtfouse, Didier Robert, and Chuanyi Wang. 2020. "Removal of Microplastics from the

Environment. A Review." *Environmental Chemistry Letters* 18:807–28. https://doi.org/10.1007/s10311-020-00983-1; Lares, Mirka, Mohamed Chaker Ncibi, Markus Sillanpää, and Mika Sillanpää. 2018. "Occurrence, Identification and Removal of Microplastic Particles and Fibers in Conventional Activated Sludge Process and Advanced MBR Technology." *Water Research* 133:236–46. https://doi.org/10.1016/j.watres.2018.01.049.

7. Simon, Matt. 2021. "People Should Drink Way More Recycled Wastewater." *Wired*. https://www.wired.com/story/people-should-drink-way-more-recycled-wastewater/.
8. American Society of Civil Engineers. 2021. "Report Card for America's Infrastructure: Wastewater." https://infrastructurereportcard.org/cat-item/wastewater/.
9. Packard, Vance. 2011. *The Waste Makers*. New York: Ig Publishing.
10. REI. 2022. "What Is Organically Grown Cotton?" https://www.rei.com/learn/expert-advice/organically-grown-cotton.html.
11. Al Jazeera. 2021. "Chile's Desert Dumping Ground for Fast Fashion Leftovers." https://www.aljazeera.com/gallery/2021/11/8/chiles-desert-dumping-ground-for-fast-fashion-leftovers.
12. Besser, Linton. 2021. "Dead White Man's Clothes." *Foreign Correspondent*. https://www.abc.net.au/news/2021-08-12/fast-fashion-turning-parts-ghana-into-toxic-landfill/100358702.
13. McCormick, Erin, Bennett Murray, Carmela Fonbuena, Leonie Kijewski, Gökçe Saraçoğlu, Jamie Fullerton, Alastair Gee, and Charlotte Simmonds. 2019. "Where Does Your Plastic Go? Global Investigation Reveals America's Dirty Secret." *Guardian*. https://www.theguardian.com/us-news/2019/jun/17/recycled-plastic-america-global-crisis.
14. Katz, Cheryl. 2019. "Piling Up: How China's Ban on Importing Waste Has Stalled Global Recycling." Yale Environment 360. https://e360.yale.edu/features/piling-up-how-chinas-ban-on-importing-waste-has-stalled-global-recycling.
15. Wan, Yong, Xin Chen, Qian Liu, Hongjuan Hu, Chenxi Wu, and Qiang Xue. 2022. "Informal Landfill Contributes to the Pollution of Microplastics in the Surrounding Environment." *Environmental Pollution* 293:118586. https://doi.org/10.1016/j.envpol.2021.118586.
16. INTERPOL. 2020. "Strategic Analysis Report: Emerging Criminal Trends in the Global Plastic Waste Market Since January 2018." https://www.interpol.int/en/News-and-Events/News/2020/INTERPOL-report-alerts-to-sharp-rise-in-plastic-waste-crime.
17. Simon, Matt. 2020. "Should Governments Slap a Tax on Plastic?"

*Wired.* https://www.wired.com/story/should-governments-slap-a-tax-on-plastic/.

18. Rosengren, Cole. 2020. "California Plastics Tax Ballot Initiative on Track for 2022 Following Court Ruling." *Waste Drive.* https://www.wastedive.com/news/recology-funding-california-plastics-tax-ballot-initiative/570717/.
19. Ford, Helen V., Nia H. Jones, Andrew J. Davies, Brendan J. Godley, Jenna R. Jambeck, Imogen E. Napper, Coleen C. Suckling, et al. 2022. "The Fundamental Links between Climate Change and Marine Plastic Pollution." *Science of the Total Environment* 806 (1): 150392. https://doi.org/10.1016/j.scitotenv.2021.150392.
20. Simon, Matt. 2019. "New IPCC Report Shows How Our Abuse of Land Drives Climate Change." *Wired.* https://www.wired.com/story/ipcc-land-report/.
21. Brizga, Janis, Klaus Hubacek, and Kuishuang Feng. 2020. "The Unintended Side Effects of Bioplastics: Carbon, Land, and Water Footprints." *One Earth* 3 (1): 45–53. https://doi.org/10.1016/j.oneear.2020.06.016.
22. Spierling, Sebastian, Eva Knüpffer, Hannah Behnsen, Marina Mudersbach, Hannes Krieg, Sally Springer, Stefan Albrecht, Christoph Herrmann, and Hans-Josef Endres. 2018. "Bio-Based Plastics: A Review of Environmental, Social and Economic Impact Assessments." *Journal of Cleaner Production* 185:476–91. https://doi.org/10.1016/j.jclepro.2018.03.014.
23. Zheng, Jiajia, and Sangwon Suh. 2019. "Strategies to Reduce the Global Carbon Footprint of Plastics." *Nature Climate Change* 9:374–78. https://doi.org/10.1038/s41558-019-0459-z.
24. Discovery Channel. 2019. "The Story of Plastic." https://www.storyofstuff.org/movies/the-story-of-plastic-documentary-film/.
25. Lewis, Simon L., and Mark A. Maslin. 2018. *The Human Planet: How We Created the Anthropocene.* New Haven, CT: Yale University Press.
26. Gowdy, John, and Lisi Krall. 2013. "The Ultrasocial Origin of the Anthropocene." *Ecological Economics* 95:137–47. https://doi.org/10.1016/j.ecolecon.2013.08.006.
27. Zalasiewicz, Jan, Colin N. Waters, Juliana A. Ivar do Sul, Patricia L. Corcoran, Anthony D. Barnosky, Alejandro Cearreta, Matt Edgeworth, et al. 2016. "The Geological Cycle of Plastics and Their Use as a Stratigraphic Indicator of the Anthropocene." *Anthropocene* 13:4–17. https://doi.org/10.1016/j.ancene.2016.01.002.
28. Wright, Ronald. 2004. *A Short History of Progress.* Toronto: House of Anansi Press.

# About the Author

Matt Simon is a science journalist at *Wired* magazine, where he covers the environment, biology, and robotics. He's the author of two previous books, *Plight of the Living Dead: What Real-Life Zombies Reveal about Our World—and Ourselves* and *The Wasp That Brainwashed the Caterpillar: Evolution's Most Unbelievable Solutions to Life's Biggest Problems*. He enjoys long walks on the beach and trying not to think about all the microplastics there.

# Index